ETHEREUM

Everything You Need to Know About It's Trade and Investment

(Best Strategies for Investing and Profiting From Ethereum)

Julie Goulart

Published by Tomas Edwards

Julie Goulart

All Rights Reserved

Ethereum: Everything You Need to Know About It's Trade and Investment (Best Strategies for Investing and Profiting From Ethereum)

ISBN 978-1-990373-69-5

All rights reserved. No part of this guide may be reproduced in any form without permission in writing from the publisher except in the case of brief quotations embodied in critical articles or reviews.

Legal & Disclaimer

The information contained in this book is not designed to replace or take the place of any form of medicine or professional medical advice. The information in this book has been provided for educational and entertainment purposes only.

The information contained in this book has been compiled from sources deemed reliable, and it is accurate to the best of the Author's knowledge; however, the Author cannot guarantee its accuracy and validity and cannot be held liable for any errors or omissions. Changes are periodically made to this book. You must consult your doctor or get professional medical advice before using any of the suggested remedies, techniques, or information in this book.

Upon using the information contained in this book, you agree to hold harmless the Author from and against any damages, costs, and expenses, including any legal fees potentially resulting from the application of any of the information provided by this guide. This disclaimer applies to any damages or injury caused by the use and application, whether directly or indirectly, of any advice or information presented, whether for breach of contract, tort, negligence, personal injury, criminal intent, or under any other cause of action.

You agree to accept all risks of using the information presented inside this book. You need to consult a professional medical practitioner in order to ensure you are both able and healthy enough to participate in this program.

Table of Contents

INTRODUCTION 1

CHAPTER 1: WHAT IS ETHEREUM? 4

CHAPTER 2: WHAT IS BLOCKCHAIN? 15

CHAPTER 3: HOW ETHEREUM DIFFERS FROM BITCOIN ... 21

CHAPTER 4: HOW ETHEREUM WORKS 27

CHAPTER 5: HOW DOES IT WORK? 33

CHAPTER 6: THE FINANCIAL HISTORY OF ETHEREUM AND ETHER 44

CHAPTER 7: HISTORY OF CRYPTOCURRENCY 49

CHAPTER 8: WHAT ARE THE DIFFERENCES BETWEEN BITCOIN AND ETHEREUM 54

CHAPTER 9: WHAT IS BLOCKCHAIN? 68

CHAPTER 10: OVERVIEW: CRYPTOCURRENCY ONLINE MONEY 74

CHAPTER 11: CAN ETHEREUM BECOME THE NEW BITCOIN? 90

CHAPTER 12: HOW TO GET ETHEREUM COINS 107

CHAPTER 13: UNDERSTANDING GAS 115

CHAPTER 14: HOW TO INVEST IN ETHEREUM (AND IS IT TOO LATE) 123

CHAPTER 15: ETHER AND OTHER CRYPTO-CURRENCIES 132

CHAPTER 16: DAO AND ETHEREUM CLASSIC 138

CHAPTER 17: BUYING AND USING ETHEREUM 143

CHAPTER 18: PROGRAMING IN ETHEREUM................... 149

CHAPTER 19: ETHEREUM SMART CONTRACTS............... 158

CHAPTER 20: RISKS AND MONEY MANAGEMENT.......... 178

CHAPTER 21: INVESTING IN ETHEREUM 186

CHAPTER 22: SHOULD I INVEST IN ETHEREUM?............ 194

CONCLUSION.. 203

Introduction

Welcome to the world of cryptocurrency. You're curious about investing and have learned about Ethereum or Bitcoin. Perhaps you know the exceptional returns possible for investments in these cryptocurrencies, which have given a massive return on investment (ROI) for those who dared to purchase them at their lows. For example, at one point the Bitcoin (BTC) to US Dollar (USD) or BTC/USD was under $1 while now it currently sits at the $600 range. That is over a 60,000% increase in any initial investment. Imagine putting $1000 into BTC at its low – you would have $600,000 now! If you weren't aware of those numbers, I might have your attention now. This book isn't about Bitcoin, but it's impossible to speak about Ethereum without explaining Bitcoin which was the first major "cryptocurrency" used. It's not that old; Bitcoin was born early in the year 2009. It set out to introduce a novel idea in a scholarly article written by the elusive

Satoshi Nakamoto. Bitcoin and all digital cryptocurrencies offer an extensive list of benefits to our current currencies such as lower transaction fees compared to traditional online payment mechanisms and banking. Cryptocurrencies are operated by a decentralized authority, unlike government and federally issued currencies. You'll never hold a physical Bitcoin or Ethereum coin in your hand since they do not exist. Rather, the currencies are all held in a virtual field of mathematics of changing balances from private or public "keys" which are synonymous with account numbers.

In this book, I hope to inform you of the reasons to invest in Ethereum and not just because of the potential ROI, but also the benefits of cryptocurrencies in themselves. I will also go over the risks, the obstacles, and major changes in Ethereum. There are over 1,000 cryptocurrencies that currently exist (albeit ~90% are just gimmicks and schemes) so it's important to choose wisely and understand everything you can if you're going to be putting real money into the blockchain. When you're done with this book, you will have both a broader and more detailed understanding of everything Ethereum and beyond.

Chapter 1: What Is Ethereum?

Ethereum can be defined as an open software platform consisting of a peer-to-peer network of virtual machines based on the blockchain technology. The platform supports developers who wish to develop and run distributed applications.

Developers can deploy decentralized or distributed applications programs (Dapps) that operate without fraud, control, interference from a third party or downtime. Ethereum is both a digital platform and a programming language that runs on a blockchain.

It supports developers to build and publish a wide range of application programs. The Ethereum network of virtual machines is optimized to execute the decentralized application programs when specific conditions are met similar to the execution of contracts. Ethereum has its decentralized blockchain that uses cryptography technology to run, store, and preserve these contracts.

Ether: On the Ethereum network, there is a token known as ether. All applications that run on the Ethereum network and its blockchain make use of ether. Ether is a cryptograph token that is essential for any application running on Ethereum. It can be described as a vehicle for navigating on the Ethereum platform so any developer who comes up with Dapps or distributed applications will need to acquire ether.

Ether is applied to the network to achieve two broad purposes. The first is to act as a currency or medium of exchange while the second is to run applications on the Ethereum blockchain. Ether is a cryptocurrency that is used by developers to pay for services and transactions within the network.

Exploring Ethereum

Ethereum can be used with any other project. One of the current projects, which is also the largest on the Ethereum platform, is the partnership between Microsoft and ConsenSys. This partnership has seen ConsenSys provide Ethereum Blockchain service on Microsoft Azure.

This enables developers, and even clients can have access to a cloud-based blockchain development by a single click.

Ethereum acts both as a decentralized app store and the Internet since it supports a new type of application. No one owns the system, but even then, it is not free to users. All users need ethers to run their applications. Ethers used in the system are more like cash or an asset. They fuel the apps within the network.

Ethereum provides what is known as the Ethereum Virtual Machine. It is known as a decentralized Turing complete machine. This is the machine that executes scripts or application programs on its vast network of public nodes. Each node needs to download and implement the virtual machine.

When developers run their programs on the system, they pay for "gas" using ethers. Gas refers to the internal pricing system that applies to the system. It allocates resources and prevents spam. Participating nodes are compensated using ethers.

Origins of Ethereum

Vitalik Buterin is a cryptocurrency programmer and research expert. He had been a developer at another blockchain company, Bitcoin. His goal for developing Ethereum was to provide a platform for developing decentralized applications. According to Vitalik, Bitcoin was in need of a scripting language which can be used to develop applications. This is why he left Bitcoin to create Ethereum.

In late 2013, he wrote a whitepaper proposing Ethereum. Shortly after the release of the whitepaper, funds for its development were raised through an online crowdsale. The crowdsale took place between July and August of 2014 and raised significant funds. Users purchased 60 million ethers while about 12 million was set aside for programmers and developers.

Ethereum got online in July of 2015. Upon going live, 11.9 million ethers were made available for the crowdsale. These account for about 13% of the total ethers supply. The system has miners who work to generate new ethers. For their work, they earn ethers which are a type of cryptocurrency.

DAO – Decentralized Autonomous Organization

In 2016, Ethereum split into two forks. The split was caused by the collapse of a project known as DAO. One fork is known as Ethereum (ETH) while the other is referred to as Ethereum Classic (ETC). The process of splitting or forking out is referred to as a hard fork. The Ethereum hard fork has created a rivalry between the two resulting systems.

The DAO event refers to an organization that was formed in 2016. Decentralized Autonomous Organization or DAO consisted of smart contracts that were developed on Ethereum. These contracts were presented at a crowdsale where they raised $150 million needed to develop the project. However, in June, the system was compromised, and thieves managed to get in and steal $50 million worth of ether. It is this theft that caused the hard fork of the system.

The Ethereum virtual machine

EVM or Ethereum virtual machine refers to the environment provided for carrying out smart contracts. The EVM is described in greater detail in Ethereum Yellow Paper written by Gavin Wood. This developers' platform is isolated from the rest of the system including the file system, the network, and all other processes.

The computers within the network, referred to as nodes, have to run the EMV. The virtual machines are implemented in some modern programming languages including Java, C++, Ruby, and Python among others.

Smart contracts

A smart contract refers merely to a computer code running on the Ethereum platform that can facilitate and promote the exchange of tangible items with a value such as property, content, shares or money. The purpose of smart contracts is to provide an exchange mechanism that enables transactions between two agents. The contracts facilitate, enforce and verify transactions and also circumvent collusion, censorship, and risks between parties.

On the Ethereum blockchain, smart contracts act like decentralized applications stored within it for execution at a later time by the Ethereum virtual machines. The contracts can be executed through a variety of programming languages and are paid for using ethers, or gas.

The smart contracts are publicly stored on each node of the Ethereum blockchain. The only challenge is that lower speeds espouse the system because each node within the system has to compute all the smart contracts and in real time. There are plenty of blockchains out there that can process code sufficiently. The problem is that most are limited in many different ways. Ethereum, however, does not face some of these limitations in its operations. It provides developers with the ability to create hundreds of applications that can achieve amazing feats.

Benefits of Ethereum

Ethereum provides a myriad of benefits to its users. It primarily enables programmers to develop and deploy a variety of decentralized applications. Dapps or decentralized applications are developed using code that operates on the blockchain so they are not regulated by any authority or individual.

The system allows any centralized program to be decentralized. There are plenty of services currently centralized that can be decentralized. These include elections management, land title registries, loans and credit from financial institutions, automobile registration, and so many others.

One major benefit of the Ethereum system is that it is immutable and no person can change the data or any code or information within the system. Another benefit is that the entire system is tamper proof and cannot be corrupted. The application programs designed for the Ethereum platform are based on a principle of consensus within the network.

The system is very secure. It is secured using the latest cryptograph technology, and there is no central point of weakness or failure. Such a secure environment is ideal for running apps which are protected against fraud and hacking attacks. Also, the system experiences zero downtime. The application programs can never be switched off by anyone, and they are never down.

Basics of cryptocurrencies

Cryptocurrencies make it very easy to send and receive funds within the network. It is much cheaper, faster and more convenient to use cryptocurrencies than traditional money transfer methods like wire transfers and so on. The cost saving is among the most significant aspects of this. Financial institutions such as banks often charge a lot of money to send money, especially international wire transfers.

Another cryptocurrency benefit is the fact that personal information is not necessary to transact. Anyone can transact using cryptocurrencies without the need to reveal their identity. Funds can be sent to recipients without the need for a bank account. Also, cryptocurrency platforms do not deny services to anyone and never discriminate. Anyone can use the digital currencies and no account can be shut down or funds impounded. This totally contrasts what banks are capable of doing.

Blockchain

The blockchain is central to the success of cryptocurrencies. It is an integral component of any digital currency as it contains a record of all transactions that occur within the network. The online ledger is updated continually and users can view transactions from their devices. This kind of decentralization and openness is unheard of in the finance world. It is hoped that in future financial institutions and other organizations will adopt a more open system, probably a blockchain-based system to ensure more openness in their operations.

A lot of experts are of the opinion that the blockchain holds potential in many different fields. It could be applied in fields such as processing financial transactions, recording stock market transactions, and in many others. Major financial institutions such as JP Morgan are of the view that a blockchain application could save firms billions in transaction costs.

Chapter 2: What Is Blockchain?

The blockchain is similar to a massive Excel-like spreadsheet, distributed over a network of computers called nodes, each node with a copy of the spreadsheet. During each hour, there are frequent updates of this ledger; as a consequence, the ledger always possesses correct information. In many ways the blockchain is a whole lot of massive, distributed, interacting databases called nodes. It is the distributed nature of the blockchain that makes it so valuable. Interactions occur using what are called protocols. These are rules written into the software.

One of the benefits is that there is never centralization of records; the records are public and on any node. It is impossible for hackers to direct focus attacks at one point, which is their usual modus operandi. A hacker would have to access to the complete network to fulfill their nefarious schemes, and that is not possible.

Proceeding with the giant spreadsheet analogy is very useful in understanding this technology. Normally, when we use a spreadsheet it is unavailable for others; they are locked out. With Google Sheets of Google Docs this locking out does not occur, allowing others to work on the same sheet, at the same time that we are working on it. The blockchain is similar except that the spreadsheet is not centralized and under the control of one organization such as Google, or Alphabet as they are now called.

What is blockchain? The blockchain is a real time computing application distributed over a network, often described as peer-to-peer. Blockchain uses cryptography and digital signatures to prove identity and enforce read/write privileges. With its data spread over many nodes, the validity of the data is far more secure from hackers than with centralized storage.

We can compare the blockchain to the Internet. It possesses both flexibility and toughness. It has identical blocks of information stored throughout the nodes on the network; there is no point in the network where failure would bring the network to its knees.

No further proof of this required than the success of Bitcoin. It has been in more or less continual operation since 2009 and has had no significant disruption. Such troubles as having occurred, are the result of human stupidity, incompetence and error. Blockchain technology has been thoroughly tested and proved. It is not an experimental technology although some of the applications built with it are.

Let us continue this comparison between the Internet and the blockchain. The Internet came to life, near the end of the 1960s at the height of the Cold War, so that it could survive a nuclear attack. It was only for military purposes. However, its use spread to universities and government. Finally, during the 1990s, it became commercially available. In spite of almost 30 years of uninterrupted use, the Internet has worked marvelously. The blockchain is in the same situation now that the Internet was in the 1990s. Many predict that its effects will be even more revolutionary than that of the Internet.

What the nodes of a blockchain are: As it is so important, we repeat that the nodes of a blockchain are the computers with the databases containing the ledger. The nodes transmit through the network details of any changes occurring in them. The changes are recorded in the blockchain. All nodes verify any transaction. The verification by one node is entirely independent of the audit by any other node. It is important to remember that the whole blockchain is updated every few minutes. No large centralized computer controls it; the updating process automatically occurs throughout the blockchain network.

The blocks of blockchain? The writing of new transactions to the blockchain takes during mining. We use the word blockchain, as there is a chain of blocks containing the transactions. Mining, and thus the manufacture of the blocks, needs very powerful computers. The miners make new coins in conjunction with making the blocks.

As mining takes place, an agreement is reached automatically between nodes. A transaction is written into the blockchain when a majority of nodes agree on the transaction's validity. Earlier it was mentioned that the process of mining requires the solution of very complex problems in mathematics. The nodes that do this are called the miners. We repeat that transactions are recorded in parts of the ledger called blocks.

A block is a table that has a timestamp; a reference, named the height, to the previously mined block, the transactions recorded in the block and the problem that miners had to solve. Anyone can see the Ethereum blockchain, as it is modified, by visiting the site etherscan.io/blocks. With this site, it is possible to see the information contained in the Ethereum blockchain. A click on any number from the Height column reveals a lot of information about that block's contents.

Doing this shows the word hash as the name of the fourth row. This word is of great importance as it is the result of a hash function. A hash function transforms a transaction to a 256-bit number, which is called a hash number and is then placed in a block after the nodes validate the hash.

The mining process assigns a unique hash number to each transaction. The chance of any two transactions with the same hash number is infinitesimally small as the number of possible hashes is so vast. Such a process is called collision resistant.

Chapter 3: How Ethereum Differs From Bitcoin

Anyone thinking about investing or trading in Ethereum has probably already seen a lot of hype about Bitcoin but not so much when it comes to Ethereum. Both currencies promise to be the wave of the future when it comes to our economy. For that reason, millions of dollars are being poured into these currencies at record-breaking rates, and there is no sign that either of them is going to be slowing down anytime soon

While both of these rely on Blockchain Technology to operate, that is where their similarities end. One of the main differences between the two is in their underlying purpose for coming into being. In the last chapter, we explained the most fundamental difference in the use of Smart Contracts, but those differences do not stop there.

Most people like to think of Bitcoin as simply digital money. It can be used to purchase products or services, to send money to other people or it can be saved just like you would with any physical currency. The primary difference between Bitcoin and regular currency comes down to its lack of formal regulations to control its worth. The value of a Bitcoin is determined not by some higher authority declaring its worth but instead by the basic principle of supply and demand. The more people want Bitcoin, the higher its value.

Today though, people are now purchasing Bitcoin and converting them into something called "tokens," which are used by many companies when they issue an Initial Coin Offering (ICO). By doing this, they can use their tokens to invest in any company they want as long as it has a cryptocoin available. These tokens work similar to buying a share in a company with its Initial Public Offering (IPO). The fluctuation in the price of these tokens, however, is determined by a completely separate market than that of Bitcoin.

People choose to buy Bitcoin for different reasons. Some want to have another tool for saving their money; one that does not require heavy regulations and monitoring like most other financial institutions. Others might choose to buy it as an investment tool, with the intention of holding onto it for a few months or a few years waiting for the price to significantly grow thus earning them a profit. And others may use Bitcoin as a means of getting into the ground floor of an ICO of a potential investment tool. Since a position in many of those companies can only be gained with a token purchase, this would be the only way to get a chance at earning any profits as they grow.

Ethereum, on the other hand, offers a whole new way of making money that Bitcoin cannot even come close to. Many are predicting that in a very short while, Ethereum is going to overtake Bitcoin as a cryptocurrency leader in this new economic structure.

Because Bitcoin was the first digital currency to be introduced, more people are familiar with it. As a matter of fact, all other "coins" are viewed based on how they compare to it. For example, take the Litecoin, a currency that has many of the same capabilities as Bitcoin (it has its own market, its own reputation, and its own established value) it is not priced anywhere near that of Bitcoin. While Bitcoin is not valued at $15,000+ at the time of this writing, Litecoin's value is just under $300.00. For now, Litecoin has a very strong standing in the digital currency world but is a long way from taking over as a market leader even with the same qualities and features as Bitcoin.

But Ethereum is different not necessarily in function but in its technology. It is not just another cryptocurrency. Its "Ether" can be bought and sold the same way as Bitcoin, it can also be used to invest in ICOs, but its power stretches far beyond that. Ethereum continues where Bitcoin stops; it is more than a currency, its technology allows its users to create new programs that can perform more than just a trade or a purchase. That makes Ethereum more of a tool for the future. The ability to create Smart Contracts means that Ethereum can adapt to a much wider range of new functions that may not have even been thought about yet.

Ethereum also has a great deal of support for its Enterprise Ethereum Alliance. A collection of Fortune 500 companies that have already come together to brainstorm even more ideas that can be applied to its advanced technology. With the use of Smart Contracts, more and more businesses are going to be keen to tap into this new technology. When details of a transaction can be coded right into the blockchain, many more functions will be performed without the use of the middleman to make sure that everything is done right. Even extremely complex actions will be performed quickly and easily saving both parties lots of money without having to shell out heavy transactions fees for every action under the sun.

While there are many technical aspects that separate Bitcoin from Ethereum, the main difference is their purpose. Bitcoin's purpose has always been as an alternative to money. It can be used as a form of payment and a means of storing value. Ethereum's main focus, however, is to make it possible for users to create peer-to-peer contracts and applications that can apply to a wide range of opportunities. They are both digital currencies, and they both use innovative Blockchain Technology, but Ether is not used as an alternative form of payment but more as a means of putting more power in the hands of users by allowing them to develop decentralized applications that are more suited to the workings of their business. Ethereum, therefore, is a more advanced network established using Blockchain Technology.

Chapter 4: How Ethereum Works

In terms of structure, the Ethereum blockchain is quite similar to that of Bitcoin. Both have the capacity to record the whole transaction history, and each node in the network can keep a copy of the transactions.

The main difference with the Ethereum network is that the nodes are stored in the most recent state of every smart contract on top of all Ether transactions. The platform has to monitor the status of the current data for all the applications, which includes the balance of the user, the smart contract codes, and the storage location.

When it comes to Bitcoin, the platform is using unspent transaction outputs to monitor who has how much digital currency. Even though it may seem complicated, the concept is actually quite easy to understand. Each time there is a Bitcoin transaction, the network will break the total amount as if it is fiat money issuing back Bitcoins in a manner that will make the information behave similarly to actual change.

In making future transactions, the Bitcoin platform should add up all your changes that are categorized as either unspent or spent. On the other hand, the Ethereum network uses accounts.

Similar to bank accounts, Ether tokens will appear in a digital wallet, and could be transferred into another account. The funds can be easily accessed, although the concept of continued relationship is non-existent.

The Ethereum Virtual Machine

In the Ethereum platform, each time an application is used, the network of thousands of computers will take charge of the processing. The contracts that are written in a specified smart contract programming language are compiled into the bytecode, which features the Ethereum Virtual Machine (EVM) that has the ability to execute and read.

All Ethereum nodes execute the contract using EVMs. Take note that each node in the network holds a duplicate of the transaction and smart contract record of the platform aside from monitoring the present state. Each time a user performs an action, the network nodes will agree that change has occurred.

The objective here is for the network of nodes and miners to take charge of the transfer instead of relying on third-party accounts such as banks or PayPal. Bitcoin miners will confirm the transfer of ownership of Bitcoin from one party to another, and the EVM will execute the contract with any rules that the developer will program in the onset.

Actual calculation on the EVM can be achieved via a machine readable language. However, developers could write smart contracts in high-level languages (such as Serpent and Solidify) that are easier for people to understand.

Smart Contracts

As with most concepts in the blockchain sector, most people are also confused over smart contracts, which is an emerging technology made possible through public blockchains. It can be difficult to understand mainly because the term doesn't clearly describe the core interaction.

Even though a basic contract will define the parameters of the relationship (typically legally binding), a smart contract will enforce a relationship with cryptographic code. Smart contracts are designed with special programs that specifically execute as defined.

First developed in 1993, the concept was originally explored by Nick Szabo – a cryptographer and computer scientist. He viewed smart contracts as some sort of digital vending machine. He described how users can input value or data and the machine will deliver a specified item.

Ethereum users could send one Ether token to anyone in the network anytime through a smart contract. In this example, the user will create an agreement, defining the specifics of the data to the contract so that it can execute specific commands.

Basically, Ethereum serves as a platform that is specifically designed for establishing smart contracts.

However, these new tools are not designed to be used as standalone as they can also compose the essential elements of decentralized applications (dApps) and even entire decentralized autonomous organizations (DAOs).

It is interesting to note that the Bitcoin network was the first platform to support basic smart contracts in the sense that the network could transmit value from one party to another. The node network could only validate transactions if specific conditions are met. However, Bitcoin is quite limited in the currency use case.

In contrast, Ethereum replaces the more restrictive design of Bitcoin and instead uses a programming language, which enables anyone with the right skillset to write their own applications. The network enables developers to program their own smart contracts or also known as autonomous agents. The programming language is regarded as a Turing complete, which means it can support a wider range of instructions and calculations.

Smart contracts can also serve as accounts for multi-signature so that the funds will only be spent if consensus has been reached. It can also take care of agreements between users and store details relevant to the app like membership records or domain registration.

Furthermore, smart contracts are likely to require assistance from other individual smart contracts. If someone will place a standard bet to project the temperature during a summer day, it could trigger a series of contracts under the hood. A contract will use external data to figure out the weather, and another contract will settle the bet according to the information it has received from the first contract once the conditions are met.

Running every contract will require transaction costs in form of Ether tokens that is largely based on the level of computational capacity required. The platform will execute the smart contracts using bytecode or a sequence of 1s and 0s, which could be read by the platform.

Chapter 5: How Does It Work?

Understanding the basics is easy enough but understanding how it works requires looking a little more deeply into the inner workings of Ethereum. One of the most common questions one might ask is how can you be sure that the automated system is working correctly to ensure that all transactions are performed as they should be. This is a very reasonable question. After all, banks go to great lengths to ensure that the people they use to verify the transactions you perform are not only well trained and qualified, but they also must be trustworthy.

The Ethereum Blockchain is made up of thousands of nodes (individual computer systems) located throughout the globe. When a transaction is submitted to the network, it is picked up and dropped into a pool where a miner will pick it up. The miner uses Ethereum's software to verify the validity of the transaction. To ensure that there is no chance of tampering or fraud, each miner is given access to the transactions at random.

Once the miner receives the transaction, he must apply a complex set of algorithms to verify their validity. This process may require a great deal of computing energy to complete. If the miner is successful in validating the transaction, it is added to a block (a compilation of all transactions made within a specific time frame); each completed block is added to the Blockchain. Each block has to connect to the block proceeding it, literally creating a chain. To form a block, there is a very specific pattern of rules that must be met.

Of course, that is the simplest explanation of how Ethereum works. However, the technology has quite a few moving parts that anyone considering investing in should come to understand and appreciate.

With every Ethereum application, the Blockchain keeps track of the most recent adjustment to any contract or document in the system. Each time a transaction is made, the data related to that account is updated and dispensed to all the nodes on the Blockchain.

The Ethereum Virtual Machine

Each time the program is used, the verified data is distributed to thousands of nodes in the network. Any contract written using Ethereum's smart contract programming language is then collected into a "byte code," which is read and executed by the Ethereum Virtual Machine (EVM).

Because the EVM is the part of the system that ensures security and the execution of all the contracts, it is extremely important to the success of Ethereum. Its focus is to 1) prevent Denial-of-service attacks and 2) ensure communication within the system can be completed without interference.

The technology behind the EVM is capable of automatically fulfilling specific tasks set out in the smart contracts. In short, the EVM is the automated system that self-executes stipulations encoded in the smart contracts. For example, if a smart contract is encoded with instructions to transfer title of a property to another person after a certain number of payments are met, it is the EVM that will automatically do that once the terms of the agreement have been complied with.

Smart Contracts

To fully grasp the concept behind smart contracts, we must go back to the Blockchain. One of the first things we need to realize is that we are not speaking of contracts in the same sense of the contracts that we conjure up in our regular business dealings. While the end result will be the same, we want to think of these smart contracts more as taking the place of our lawyers, accountants, and notaries.

The main idea behind them is that they must serve as a self-executing mechanism used to fulfill any agreements between two parties. They are specially written codes that perform specific tasks when they are told. So, a smart contract is actually a computer programming code that tells the EVM when to perform a certain task.

By implementing smart contracts to the Blockchain, it fosters a field of trust. This is one of the reasons why the miner's role is so important in the actual validation of every transaction that passes through Ethereum's network.

Swarm and Whisper

As you might imagine, storing all of this data would take up a lot of space. Being able to store code that consists of numbers and digits as with Bitcoin's simple platform is a far cry from the detailed data that must be kept in Ethereum's database. Their solution to this problem comes in the form of Swarm, a decentralized storage branch in the network. Swarm's role is to function in the same way as Dropbox does, by providing a platform for users to store their data and share information. This will all be managed and controlled with a consensus algorithm that will verify not just the content but also to clear a path for those authorized to use it.

The advantage of this type of storage system is again its decentralization. Without any one single entity responsible for maintaining the data, there are no possible means for any individual or group to remove it from the system.

To that end, the future for this aspect of the Ethereum network is without end. Already, many are considering it as a place to archive records, historical documents, medical records, and other forms of data that could have immense value to succeeding generations.

Of course, this new storage system is in its early stages and is not expected to launch until the summer of 2018 along with a new consensus algorithm miners will use to verify data. As expected, the team behind Swarm's development is already at work reworking the entire network to incorporate not just its storage capabilities but also the synchronization and retrievability of all the data it is expected to store.

Since the new storage platform will be using the new proof-of-concept algorithm, it is expected to be completely compatible with other tools used in the Ethereum network.

With Swarm, users will be able to store medical records, property deeds, and any other type of document by using nodes to connect to the swarm network. Users will receive a swarm address that they can send to the other party in the agreement allowing them to access the documentation.

Whisper, on the other hand, another technology that is utilized by Ethereum is a communication protocol that allows Dapps to interact with each other. The Whisper protocol allows for peer-to-peer messaging to be passed along the network.

It is a system that will operate completely separate from the Blockchain but in conjunction with it. Whisper has a number of characteristics that will eventually allow it to become a standard feature for developers when they want to add a specific message to anything they are coding.

Whisper encrypts the messages in the same way that every other aspect of a transaction will be encrypted to ensure that not only are the contents preserved and protected but also the metadata as well.

This can serve as a huge plus for the users and for the developers. By not having free access to data, developers won't be held as liable for its protection. They cannot accidentally "leak" information nor can they be coerced into divulging it to any unauthorized individual. One can think of it as an added layer of security for everyone concerned.

Sometimes referred to as the Holy Trinity, these three parts of the system (Ethereum, Swarm, and Whisper) when combined can produce a formidable network that will be hard for any centralized platform to compete with. No doubt, this added level of protection when used on the Blockchain can be an appealing option for many corporations that wish to at least divert some of their business affairs away from the traditional method of centralization that we have been relying on for centuries.

What Exactly is Solidity?

The Ethereum network brings all of this together using a unique programming language called Solidity. This high-level programming language is used to instruct the EVM on the protocols and demands of each contract that will have to be self-executed in the future.

Solidity is the scripting language, which enforces the verifying and enforcing the terms of whatever agreements are made and put into the system. All of Ethereum's source code is written in Solidity and remains consistent with the execution of every agreement.

Proof of Work

The proof-of-work (PoW) algorithm is the tool by which miners will be able to ensure that each block added to the chain is the only version of the transactions contained in it. It is also the tool by which unauthorized individuals will be prevented from hacking the system and sabotaging it in some way.

With the PoW, miners work to add the next block to the chain by solving a cryptographic puzzle. The first miner to solve the puzzle gets to add the block to the chain and in reward will be paid in Ether along with a small transaction fee.

Once they receive a transaction, they run the block's header metadata through a hash function to find the hash. The resulting hash must match the current target. Once the hash matches all the requirements, it must be checked and verified by another miner before it can be included in the Blockchain.

Using this encryption method, Ethereum creates a completely new block every 12-15 seconds, much faster than Bitcoin's new block creation time of once every 10 minutes.

It is very difficult to solve the solutions to the puzzles created by the encryption process, so it is practically impossible to cheat when working out the Proof-of-work algorithm. Each hash created must not only match with all of the transactions included in the block, but it must also match the block before it. Also, each character in the code has to have some sort of relationship with the other characters in the hash. This process creates millions of variables with only one possible outcome. It is the Blockchain's system for ensuring that no data has been altered or tampered with in any way. In fact, if even one figure is changed it not only affects the hash that was created but it will also affect every hash created after it.

Once the hash is created, it is added to the Blockchain, and the entire process begins again with the miner working to solve the next block in the chain.

While you may not understand every aspect of the Ethereum network, having knowledge like this can be very beneficial in understanding how the system works and why it is considered to be a platform for the ages. Deciding if you want to be a part of this world starts with understanding exactly what's involved in the Ethereum Blockchain and why so many people are intrigued by it.

Chapter 6: The Financial History Of Ethereum And Ether

At an initial level, Ethereum was described by Vitalik Buterin in a white paper. Vitalik Buterin was a programmer working on Bitcoin Magazine in 2013 and he was working on a goal to build decentralized applications. Buterin suggested that scripting linguistic is necessary for Bitcoin application development. He failed to get agreement and decided to develop a new platform with a general scripting language.

In January 2014, the main team of Ethereum was Charles Hoskinson, Anthony Di Iorio, Mihai Alisie, and Vitalik Buterin. Formal development of this project started in 2014 through a Swiss organization called EthSuisse (Ethereum Switzerland GmbH). Afterward, a non-profit Swiss foundation (Stiftung Ethereum) Ethereum Foundation was developed. An online crowdsale had funded this development between July and August 2014. The buyers of Ethereum got ether token with bitcoin.

Architecture of Ethereum

Presale of Ether

It is a value token of the blockchain Ethereum and listed under ETH code. It is traded on the exchanges of cryptocurrency. It is utilized to pay for computational services and transaction fees on the network of Ethereum. Tokens may be volatile as per circumstances, such as plunge of ether from $21.50 - $8 on 17th June 2016, when DAO was hacked. The value of ether had increased to more than $400 in June 2017. This rise was equal to 5,000% since the commencement of the year.

The volatility of price on a single exchange may exceed the volatility on the prices of Ether token. A "flash crash" is triggered by the large selling order on an exchange dropped the cost on that exchange to $0.10 as each offer to buy was engrossed, after which the rate quickly recovered to $300 and more.

Ethereum Virtual Machine

The EVM (Ethereum Virtual Machine) is a suitable runtime atmosphere for smart contracts in Ethereum. Buterin initiated his work with Dr. Gavin Wood to discover Ethereum. Dr. Wood released Yellow Paper Ethereum to explain the specifications of EVM. As per his papers, EVM is sandboxed and completely sequestered from filesystem, network and other procedures of the host computer. Each Ethereum in its network runs a particular EVM implementation and implements the similar instructions. EVMs have been implemented in Web Assembly (under development), Rust, Ruby, Python, JavaScript, Java, Haskell, Go and C++.

Development of Ethereum

After presale of Ether, the development of Ethereum was officially announced as a non-profit organization ETH DEV. This organization managed the development of Ethereum with Vitalik Buterin and team. The whole team of three directors delivered PoC (Proof-of-Concept) released for the growing community of Ethereum. On August 2014, the Ethereum's sale reaches to more than $14 million. In September 2014, the pre-sale investors obtained 60m Ether and 12m Ether were provided to the Ethereum's development team. The left behind ether was provided to the non-profit organization. After all this, it was October 2014, when Ethereum protocol approved five ETH cream. Rising network of miners for Ethereum and increasing traffic on the forum of Ethereum clearly defined that the network is rapidly enticing developer community.

Frontier Launch

The initial version of Ethereum was launched as Frontier on 30th July 2015. Moreover, minors were really interested and they frequently join this network to secure Ethereum blockchain and earn Ether. In fact, Frontier was the initial milestone in cryptocurrency Ethereum and intended as the beta version by developers. The release of Frontier became capable and reliable solutions and resulted in the enhanced ecosystem of Ethereum. A prearranged protocol

"Homestead" was added to the network of Ethereum on 14th March 2016. In the month of May 2016, DAO set a special $150 million record and increase the popularity of Ethereum. Afterward, some unidentified group hacked DAO and claim $50 worth Ether.

Advance Developments

On second July 2016, the network of Ethereum was split in ETH (Ethereum) and ETC (Ethereum Classic). After a particular period, Ethereum debated for reorganized decision making and intentionality extra-protocol and ETC for blockchain law code and immutability. After that, Ethereum users, developers, partners, and miners left the ETC (Ethereum Classic) network on 28th of July 2016. Finally, it was November 2016 when Ethereum debloated blockchain, prevent span attacks and increase DDoS.

Chapter 7: History Of Cryptocurrency

Cryptocurrency existed as a theoretical construct long before the first digital alternative currencies debuted. Early cryptocurrency proponents shared the goal of applying cutting-edge mathematical and computer science principles to solve what they perceived as practical and political shortcomings of "traditional" fiat currencies.

Technical Foundations

Cryptocurrency's technical foundations date back to the early 1980s, when an American cryptographer named David Chaum invented a "blinding" algorithm that remains central to modern web-based encryption. The algorithm allowed for secure, unalterable information exchanges between parties, laying the groundwork for future electronic currency transfers. This was known as "blinded money."

By the late 1980s, Chaum enlisted a handful of other cryptocurrency enthusiasts in an attempt to commercialize the concept of blinded money. After relocating to the Netherlands, he founded DigiCash, a for-profit company that produced units of currency based on the blinding algorithm. Importantly, DigiCash's control wasn't decentralized, as is the case with Bitcoin and most other modern cryptocurrencies – DigiCash itself had a monopoly on supply control, similar to central banks' monopoly on fiat currencies.

DigiCash initially dealt directly with individuals, but the Netherlands' central bank cried foul andquashed this idea. Faced with an ultimatum, DigiCash agreed to sell only to licensed banks, seriously curtailing its market potential. Microsoft later approached DigiCash about a potentially lucrative partnership that would allow early Windows users to make purchases in its currency, but the two companies couldn't agree on terms, and DigiCash went belly-up in the late 1990s.

Around the same time, an accomplished software engineer named Wei Dai published a white paper on b-money, a virtual currency architecture that included many of the basic components of modern cryptocurrencies, such as complex anonymity protections and decentralization. However, b-money was never deployed as a means of exchange.

Shortly thereafter, a Chaum associate named Nick Szabo developed and released a cryptocurrency called Bit Gold, which was notable for using the block chain system that underpins most modern cryptocurrencies. However, Bit Gold never gained popular traction and is no longer used as a means of exchange.

Pre-Bitcoin Virtual Currencies

After DigiCash, much of the research and investment in electronic financial transactions shifted to more conventional, though digital, intermediaries, such as PayPal. A handful of DigiCash imitators, such as Russia's WebMoney, sprang up in other parts of the world.

In the United States, the most notable virtual currency of the late 1990s and 2000s was known as e-gold. e-gold was created and controlled by a Florida-based company of the same name. e-gold, the company, basically functioned as a digital gold buyer. Its customers, or users, sent their old jewelry, trinkets, and coins to e-gold's warehouse, receiving digital "e-gold" — units of currency denominated in ounces of gold. e-gold users could then trade their holdings with other users, cash out for physical gold, or exchange their e-gold for U.S. dollars.

At its peak in the mid-2000s, e-gold had millions of active accounts and processed billions of dollars in transactions annually. Unfortunately, e-gold's relatively lax security protocols made it a popular target for hackers and phishing scammers, leaving its users vulnerable to financial loss. And by the mid-2000s, much of e-gold's transaction activity was legally dubious — its laid-back legal compliance policies made it attractive to money laundering operations and small-scale Ponzi schemes. The platform faced growing legal pressure during the mid- and late-2000s, and finally ceased to operate in 2009.

Chapter 8: What Are The Differences Between Bitcoin And Ethereum

Cryptocurrency has changed a lot over the past few years. What started out as a simple idea to purchase items online has grown all over the world. People use bitcoins to purchase items while keeping their identities safe (much better than using your credit card and keeping all that personal information out there), but there are other ways to use this as well. Cryptocurrency has grown to include investing, selling items, making an income, and even backing up new start-up businesses. It is a whole new world of currency and many people want to learn how to take advantage of it.

Cryptocurrency, even though it is a newer idea, is still a form of money. For some people, that is all they need to know about it. There are a few stores who will accept cryptocurrency (it can be converted to American dollars as well as to some other currencies in some countries), as long as you have the right barcodes to make this happen. But for the most part, cryptocurrency is used and traded online.

Currently there are many different types of cryptocurrencies available. Bitcoin is probably one of the best known, but each of them will have their own value and it does go up and down a bit depending on the market. The Bitcoin right now is worth about $3,461 (Jul 23, 2017) American dollars, but other cryptocurrencies may be worth more or less depending on how many people are using them. These currencies are often considered stronger than regular currency, though, since they won't be tied to a bank or a government.

The biggest difference between cryptocurrencies and regular money from the bank is that these cryptocurrencies are not going to be backed by the government. The cryptocurrencies are going to have rules that were programmed into them and this will control how they work, rather than the government. Some people prefer this method because they don't have to worry about the government meddling in the value of their money.

All cryptocurrencies are going to be based on technology that includes data verification and advanced encryption, which helps to keep them secure and makes it hard for someone to forge the currency. In addition, these currencies are often designed with a specific maximum amount of currency available so inflation will stay pretty steady over time.

Now, there are going to be some differences between the cryptocurrencies, including who uses them, how they are designed, and how much currency is available, but they can often come from the same basis and will be used by consumers in the same way. This guidebook is going to focus on two of the big names in cryptocurrencies; Bitcoin and Ethereum.

The basics of Bitcoin

Let's start with Bitcoin. This is for sure the most popular cryptocurrency. It is used to pay for things, but like the other cryptocurrencies, it is not tied to a government or a bank. This type of cryptocurrency is going to work like a big ledger that all of the users are able to share. When you get paid or pay for something using the Bitcoin system, all of these transactions are going to be placed on a ledger. Computers will then compete with each other to confirm the transaction, usually by solving a math equation that is complex, before they are rewarded with more of the Bitcoin. This process is called mining and is one of the ways that people can earn a lot of Bitcoin, but we will talk more about that later.

Bitcoin was developed in 2009, but very few people heard about it or used it at the time. The idea behind developing this kind of currency was to take control of money away from governments and central bankers, two groups that often manipulate money to their own personal gain. It is still unknown who exactly developed Bitcoin. The group Satoshi Nakamoto are recognized as the creators, but no one knows who is behind this group.

So, how does Bitcoin work? To start, you need to get the Bitcoin wallet app and have it on your phone or computer. There are several ways that you can then earn Bitcoin. If you would just like to make some purchases with Bitcoin, you can use your banking information to purchase or exchange for them. Some merchants will open up the wallet and start accepting Bitcoin for their goods and services. Others will invest, give money to start-up companies, or use the mining technique from before to earn more Bitcoin.

There are quite a few places where you are able to use Bitcoin, and that list is growing as more and more people want to use this online currency. Some big online merchants, like Overstock.com and OkCupid, will accept Bitcoin. Some car dealership allows you to use Bitcoin, like one in Southern California who had a customer buy a Tesla Model S with their Bitcoins. Many people who are offering goods and services can list them online and will accept Bitcoin. If you want to purchase something with Bitcoin, you simply need to check out their website to see if this is an option.

So why is Bitcoin such a great option to work with? Some say that it is a good deal for a business, especially a small one, to accept Bitcoin. In most cases, these businesses would need to pay some sort of fee to the credit card companies each time that a customer uses a card to pay for something. This can get expensive for some companies. But with Bitcoin, the transactions are free and this could save the merchants a lot of money.

Bitcoin is also anonymous. While you will need to provide your banking information if you would like to exchange or purchase Bitcoin, all the information is encrypted and no one is able to trace it back to you. This makes it nice to protect against identity theft, though there are some people who use it to purchase items that may not be legal in their areas.

Bitcoin is also really secure. Through the process of mining, the transactions will all get a security code. Each number in the code is dependent on the next one, so if one number is changed, the rest of the code will change as well. This can help make it easier for a merchant to see if a transaction is legitimate or not before they send out their product.

And there are a lot of great companies that will let you shop with Bitcoin. Since it is not tied to a bank or a government agency, you can rest assured that the value isn't going to just suddenly drop out like it can with traditional money. This is a sigh of relief for a lot of people, especially those who may have some turmoil with the banks right now.

The split in Bitcoin

One recent change with Bitcoin is that the currency was recently split in two. Basically, the split has occurred because some of the ideological, technological, and political debates about how to grow Bitcoin have come together and forced it. One side of the split, which is known as Bitcoin Cash (BCH), is going to help scale up Bitcoin and make it easier for everyone to get a hold of it. There have been some growing pains with making this available to everyone, and some people believe that using the same software to create a brand-new currency is the best way to deal with this.

Over the years, Bitcoin has grown quite a bit and because of this, the blockchain that controls the transactions is becoming kind of congested and slowing down. For some, Bitcoin Cash seems to be the answer. BCH is novel software that will allow for eight times the amount of transactions on each block chain, helping to clear out some of the congestion that is there. However, the value of the Bitcoin in Bitcoin Cash could go down though. Bitcoin is worth about $ $3,461 each and the Bitcoin Cash would be worth around $313.70 (Jul 23, 2017).

Since these work similarly, other than the price of each will change and the process should be faster, what is this change going to mean for businesses and consumers who use the system? Anyone who already has Bitcoin is going to end up having the same amount of BCH. However, it is important to know that some of the exchanges that used Bitcoin are not going to accept Bitcoin Cash, which could be a bad thing for the potential spread of Bitcoin.

The hardest part of this process is getting businesses to start accepting it as payment. It doesn't matter so much what the changes are, though adding in more transactions to a blockchain will certainly help work out some of the kinks of Bitcoin. But if Bitcoin Cash is to become mainstream and be used for regular transactions, businesses need to be willing to accept it. This could take some time, but as more people want to use Bitcoin, it is likely that businesses will want to jump on as well.

What is Ethereum?

Ethereum is another choice that you can use in cryptocurrency. This one is a bit newer than Bitcoin so it hasn't come up with quite as big of a following as that option, but it is still popular. There are some differences between these two currencies, though. For example, unlike the Bitcoin platform, Ethereum has been designed to be a smart contract platform which is also based on the blockchain technology.

The tokens for this cryptocurrency are available on a platform that is known as Ether. The ETH platform was designed specifically to be used for making any payment for hosting and accessing applications that are on the Ethereum blockchain. These apps will run on a blockchain, which is a powerful infrastructure that can be shared globally and which will move value around while also representing the ownership of that property. This can be created because it will allow the developers to create markets, move funds through the right instructions, store their debts and promises in registries, and many other options. The best part is that all of these can be done without needing to add in a middle man.

The idea of Ethereum was released in 2014 by fans all around the world. This type of currency was developed by a company known as the Ethereum Foundation, which is a Swiss non-profit company, and there are contributions to this platform from all over the globe.

As you can see, Ethereum is a bit different and there are some limitations that come with this one that may not be seen with Bitcoin, but it is still a good option to work with. First, it is possible to work on mining inside of Ethereum as well in order to earn more of this cryptocurrency as well. If you have worked with Bitcoin in the past, you have seen how the mining can work. This process is when people who are already on the network will work on the codes that help maintain the applications and let them continue running. This process can also be used to start up new blocks on the blockchain and to go through and process the transactions that occur on the network. When the miners are successful, they will receive Ether, or the minted crypto-tokens, as their reward or payment.

Just like with the other cryptocurrencies, Ether has a value and you will be able to buy or sell them based on the current value of other types of cryptocurrencies.

There are quite a few cryptocurrencies that are happy to do exchanges with Ethereum so it is easy to find the one that is right for you.

The differences between Bitcoin and Ethereum

While both of these get their power by the principle of cryptography and distributed ledgers there are some technical differences in the two currencies respectively technologies. First of all, they differ regarding their programming languages. Ethereum is utilizing Turning while Bitcoin uses stack based languages. The block time is different as well. Ethereum is going to get done with transactions within a few seconds, but it does take longer when using Bitcoin. And the basic builds of these two currencies are different in the algorithms that they use.

But most people are not interested in these technical differences (other than the transaction time being faster is usually appreciated). The biggest difference that you are going to notice between these two is the purpose of using them. Bitcoin was designed to be an alternative to regular money. This means that you are able to use it in stores and in pretty much the same way that you do money, as long as the company accepts that form of payment. You get the added bonus of not having a bank or a government agency control the money. If you are looking for a safe, secure, and anonymous way to make purchases, Bitcoin is the right one for you.

On the other hand, Ethereum was developed as a type of platform that you would use to facilitate peer-to-peer contracts and applications with its own vehicle for currency. It is a way for people to get money to help create new applications, money that they would get from investors who are interested in the app. While Ethereum is a digital currency just like Bitcoin, the purpose is to monetize and facilitate Ethereum to work so that developers are able to build and then run distributed applications.

In summary, while both of these work online and use the same idea of a blockchain, they are not even going to compete with each other. But since the Ether has gained a lot of popularity, it has become a competitor to many of the other cryptocurrencies, especially when looking at it from a trading perspective. For example, the market cap of the Ether is higher than Litecoin or Ripple but it still has a bit to go to reach Bitcoin. With that being said, both of these cryptocurrencies could be effective and popular at the same time, since they have different end goals.

In spite of the fact that Ethereum and Bitcoin may have comparative perspectives, the two monetary standards are altogether different as to the utilization as well as the eventual fate of the two digital assets. At the same time as the fate of ETH is vague there is an appealing investment opening to catch the potential expansion of the innovation. Concerning BTC, there is still space for development along with esteem; however, it won't be at the unstable rate it encountered in the commencement years.

Chapter 9: What Is Blockchain?

Blockchain means Blockchain technology. Blockchains allow the distribution of digital information without being copied. By doing this, Blockchain has become the most important section of a new sort of Internet. Blockchain was originally designed for the use of Bitcoin, the first and most valuable digital currency. However, as with all new technologies, new uses are continually being discovered for it. Please note that for example Bitcoin and Ethereum run on different implementations of Blockchains.

Ledger

In our discussion about Blockchains, we are about to introduce a new term, which is ledger. Before proceeding, we had better define what a ledger is. A ledger is simply a collection of financial accounts and would normally contain records of assets, debts, transactions, inventory, liquidity, etc.

Blockchain is a Ledger

The Blockchain is a digital ledger of complete incorruptibility. Not only can it be programmed to keep records of financial dealings but also it can keep records of anything at all. You may well ask, "How is this possible?"

A comparison, at this stage, is very useful, involving a spreadsheet, with many thousands of duplicates spread over a vast network of computers. At regular and frequent intervals, this spreadsheet is updated so the ledger never becomes out of date. This is what the idea of Blockchain technology is. It envisages a vast distributed database. What makes this database so valuable is that it is not stored in a single location, meaning that its records are both public and easily verified. There is no single, central location for hackers to attack, corrupt and wreak havoc as they now so often do.

Let us continue the spreadsheet analogy. At present, we can work on such a spreadsheet but, in doing so, then usually other users are locked out. However, if we have our spreadsheet as a Google sheet in

Google Docs then this locking out does not need to occur and two or more users can work on the same sheet simultaneously. There are a vast number of legal documents that should have such treatment. At the moment an important transaction takes considerable time as it is passed from point to point in its development, as it is checked and signed off.

Some people viewing the spreadsheet case may feel that this creates a distributed system and may well say "Why not extend these ideas to finance and other areas? The problem is that, at the moment, it is Google that stores that spreadsheet in its database in the cloud. The company that owns Google has ultimate control. You may well say the owners of Google had as their motto 'Don't be evil,' so what is the problem?

In actual fact, Google is now Alphabet Inc. who replaced that motto by 'Do the right thing.' Despite the high and virtuous mottoes, these people still have ultimate control. Would you trust them with your money?

Here is where Blockchain technology is so useful. It is similar to the World Wide Web in that it has a built in toughness and flexibility. The storage of identical information blocks across the network means that the Blockchain cannot be controlled by a single person or organization and that there are no single points of failure.

This is neither conjecture nor idle speculation. This idea has been thoroughly proved by Bitcoin. Bitcoin was the first example of the application of Blockchain technology; it has existed since 2008 and has operated largely free of disruption in that time. Problems, arising, have done so as a result of human error and incompetence. The actual technology has proved to be blameless.

Another successful example of a distributed network is the Internet itself. Originally designed to withstand a nuclear attack, it was developed in the late 1960s for military purposes. It was used commercially for the first time in the 1990s. In nearly 30 years of use, this system as a whole has worked perfectly although some would say it has been horribly abused.

In Blockchain, the network lives in a state of agreement. It is automatically updated about every 10 minutes. This update is not carried out from some central computer. It must be seen as an inbuilt mechanism of the network that helps to protect it. The information it contains is never out of date and it is public. This last statement is very important and means there is a public record on the Blockchain of any transaction which cannot be altered.

In addition, any information on the network is incorruptible. This means that it is impossible to alter any units of information on the network without a vast amount of computing power. Some criminal organization or rogue state might do this in the false belief that they could get Bitcoins or some other unit of value. If they did this, then the Bitcoins would immediately lose their value.

In the West, people are waking up to the fact that large technology companies do not always have their best interests at heart. They are also becoming very suspicious of the hypocrisy of their governments who are forever breaking their promises and always looking for new ways to spy on them. Despite this, the citizens of countries in the West probably have some faith in the existing system.

It is in the countries of Eastern Europe, the Middle East, Latin America and many parts of Asia that Blockchain technology will find its greatest opportunities. In these countries, long experience has taught their citizens to be extremely suspicious of large organizations such as governments and international companies.

If you would like to delve deeper into "how" Blockchains generally work, there is a fantastic, but technical, video on YouTube: "Blockchain 101 - A Visual Demo" by Anders Brownworth. I personally learned most of my technical understanding from this video and highly recommend it.

Chapter 10: Overview:

Cryptocurrency Online Money

Cryptocurrencies, or virtual currencies, are digital means of exchange created and used by private individuals or groups. Because most cryptocurrencies aren't regulated by national governments, they're considered alternative currencies mediums of financial exchange that exist outside the bounds of state monetary policy.

Bitcoin is the preeminent cryptocurrency and first to be used widely. However, hundreds of cryptocurrencies exist, and more spring into being every month.

What Is Cryptocurrency?

Cryptocurrencies use cryptographic protocols, or extremely complex code systems that encrypt sensitive data transfers, to secure their units of exchange. Cryptocurrency developers build these protocols on advanced mathematics and computer engineering principles that render them virtually impossible to break, and thus to duplicate or counterfeit the protected currencies. These protocols also mask the identities of cryptocurrency users, making transactions and fund flows difficult to attribute to specific individuals or groups.

Cryptocurrencies are also marked by decentralized control. Cryptocurrencies' supply and value are controlled by the activities of their users and highly complex protocols built into their governing codes, not the conscious decisions of central banks or other regulatory authorities. In particular, the activities of miners – cryptocurrency users who leverage vast amounts of computing power to record transactions, receiving newly created cryptocurrency units and transaction fees paid by other users in return are critical to currencies' stability and smooth function.

Importantly, cryptocurrencies can be exchanged for fiat currencies in special online markets, meaning each has a variable exchange rate with major world currencies (such as the U.S. dollar, British pound, European euro, and Japanese yen).

Cryptocurrency exchanges are somewhat vulnerable to hacking and represent the most common venue for digital currency theft.

Most, but not all, cryptocurrencies are characterized by finite supply. Their source codes contain instructions outlining the precise number of units that can and will ever exist. Over time, it becomes more difficult for miners to produce cryptocurrency units, until the upper limit is reached and new currency ceases to be minted altogether. Cryptocurrencies' finite supply makes them inherently deflationary, more akin to gold and other precious metals — of which there are finite supplies — than fiat currencies, which central banks can, in theory, produce unlimited supplies of.

Due to their political independence and essentially impenetrable data security, cryptocurrency users enjoy benefits not available to users of traditional fiat currencies, such as the U.S. dollar, and the financial systems that those currencies support. For instance, whereas a government can easily freeze or even seize a bank account located in its jurisdiction, it's very difficult for it to do the same with funds held in cryptocurrency even if the holder is a citizen or legal resident.

On the other hand, cryptocurrencies come with a host of risks and drawbacks, such as illiquidity and value volatility, that don't affect many fiat currencies. Additionally, cryptocurrencies are frequently used to facilitate gray and black market transactions, so many countries view them with distrust or outright animosity. And while some proponents tout cryptocurrencies as potentially lucrative alternative investments, few (if any) serious financial professionals view them as suitable for anything other than pure speculation.

How Cryptocurrencies Work

The source codes and technical controls that support and secure cryptocurrencies are highly complex. However, laypeople are more than capable of understanding the basic concepts and becoming informed cryptocurrency users.

Functionally, most cryptocurrencies are variations on Bitcoin, the first widely used cryptocurrency. Like traditional currencies, cryptocurrencies' express value in units — for instance, you can say "I have 2.5 Bitcoin," just as you'd say, "I have $2.50." Several concepts govern cryptocurrencies' values, security, and integrity.

BLOCK CHAIN

A cryptocurrency's block chain is the master ledger that records and stores all prior transactions and activity, validating ownership of all units of the currency at any given point in time. As the record of a cryptocurrency's entire transaction history to date, a block chain has a finite length containing a finite number of transactions that increases over time.

Identical copies of the block chain are stored in every node of the cryptocurrency's software network — the network of decentralized server farms, run by computer-savvy individuals or groups of individuals known as miners, that continually record and authenticate cryptocurrency transactions.

A cryptocurrency transaction technically isn't finalized until it's added to the block chain, which usually occurs within minutes. Once the transaction is finalized, it's usually irreversible unlike traditional payment processors, such as PayPal and credit cards, most cryptocurrencies have no built-in refund or chargeback functions, though some newer cryptocurrencies have rudimentary refund features.

During the lag time between the transaction's initiation and finalization, the units aren't available for use by either party. The block chain thus prevents double-spending, or the manipulation of cryptocurrency code to allow the same currency units to be duplicated and sent to multiple recipients.

PRIVATE KEYS

Every cryptocurrency holder has a private key that authenticates their identity and allows them to exchange units. Users can make up their own private keys, which are formatted as whole numbers between 1 and 78 digits long, or use a random number generator to create one. Once they have a key, they can obtain and spend cryptocurrency. Without the key, the holder can't spend or convert their cryptocurrency rendering their holdings worthless unless and until the key is recovered.

While this is a critical security feature that reduces theft and unauthorized use, it's also draconian – losing your private key is the digital equivalent of throwing a wad of cash into a trash incinerator. While you can create another private key and start accumulating cryptocurrency again, you can't recover the holdings protected by your old, lost key.

WALLETS

Cryptocurrency users have "wallets" with unique information that confirms them as the temporary owners of their units. Whereas private keys confirm the authenticity of a cryptocurrency transaction, wallets lessen the risk of theft for units that aren't being used. Wallets used by cryptocurrency exchanges are somewhat vulnerable to hacking – for instance, Japan-based Bitcoin exchange

Mt. Gox shut down and declared bankruptcy after hackers systematically relieved it of more than $450 million in Bitcoin exchanged over its servers.

Wallets can be stored on the cloud, an internal hard drive, or an external storage device. Regardless of how a wallet is stored, at least one backup is strongly recommended. Note that backing up a wallet doesn't duplicate the actual cryptocurrency units, merely the record of their existence and current ownership.

MINERS

Miners serve as record-keepers for cryptocurrency communities, and indirect arbiters of the currencies' value. Using vast amounts of computing power, often manifested in private server farms owned by mining collectives comprised of dozens of individuals, miners use highly technical methods to verify the completeness, accuracy, and security of currencies' block chains. The scope of the operation is not unlike the search for new prime numbers, which also requires tremendous amounts of computing power.

Miners' work periodically creates new copies of the block chain, adding recent, previously unverified transactions that aren't included in any previous block chain copy – effectively completing those transactions. Each addition is known as a block. Blocks consist of all transactions executed since the last new copy of the block chain was created, usually a few minutes prior.

The term "miners" relates to the fact that miners' work literally creates wealth in the form of brand-new cryptocurrency units. In fact, every newly created block chain copy comes with a two-part monetary reward: a fixed number of newly minted ("mined") cryptocurrency units, and a variable number of existing units collected from optional transaction fees (typically less than 1% of the transaction value) paid by buyers. Thus, cryptocurrency mining is a potentially lucrative side business for those with the resources to invest in power- and hardware-intensive mining operations.

Though transaction fees don't accrue to sellers, miners are permitted to prioritize fee-loaded transactions ahead of fee-free transactions when creating new block chains, even if the fee-free transactions came first. This gives sellers an incentive to charge transaction fees, since they get paid faster by doing so, and so it's fairly common for transactions to come with fees. While it's theoretically possible for a new block chain copy's previously unverified transactions to be entirely fee-free, this almost never happens in practice.

Through instructions in their source codes, cryptocurrencies automatically adjust to the amount of mining power working to create new block chain copies copies become more difficult to create as mining power increases, and easier to create as mining power decreases. The goal is to keep the average interval between new block chain creations steady at a predetermined level for instance, Bitcoin's is 10 minutes.

BITCOIN MINING: FINITE SUPPLY

Although mining periodically produces new cryptocurrency units, most cryptocurrencies are designed to have a finite supply. Generally, this means that miners receive fewer new units per new block chain as time goes on. Eventually, miners only receive transaction fees for their work.

This has yet to happen with any extant cryptocurrency, but observers predict that the last Bitcoin unit will be mined sometime in the mid-22nd century, if current trends continue. Finite-supply cryptocurrencies are thus more similar to precious metals, like gold, than to fiat currencies – of which, theoretically, unlimited supplies exist.

CRYPTOCURRENCY EXCHANGES

Many lesser-used cryptocurrencies can only be exchanged through private, peer-to-peer transfers, meaning they're not very liquid and are hard to value relative to other currencies – both crypto- and fiat. More popular cryptocurrencies, such as Bitcoin and Ripple, trade on special secondary exchanges similar to forex exchanges for fiat currencies. (The now-defunct Mt. Gox is one example.) These platforms allow holders to exchange their cryptocurrency holdings for major fiat currencies, such as the U.S. dollar and euro, and other cryptocurrencies (including less-popular currencies). In return for their services, they take a small cut of each transaction's value – usually less than 1%.

Cryptocurrency exchanges play a valuable role in creating liquid markets for popular cryptocurrencies and setting their value relative to traditional currencies. However, exchange pricing can still be extremely volatile.

ADVANTAGES OF CRYPTOCURRENCY

Built-in Scarcity May Support Value

Most cryptocurrencies are hardwired for scarcity – the source code specifies how many units can ever exist. In this way, cryptocurrencies are more like precious metals than fiat currencies. Like precious metals, they may offer inflation protection unavailable to fiat currency users.

Loosening of Government Currency Monopolies

Cryptocurrencies offer a reliable means of exchange outside the direct control of national banks, such as the U.S. Federal Reserve and European Central Bank. This is particularly attractive to people who worry thatquantitative easing (central banks' "printing money" by purchasing government bonds) and other forms of loose monetary policy, such as near-zero inter-bank lending rates, will lead to long-term economic instability.

In the long run, many economists and political scientists expect world governments to co-opt cryptocurrency, or at least to incorporate aspects of cryptocurrency (such as built-in scarcity and authentication protocols) into fiat currencies. This could potentially satisfy some cryptocurrency proponents' worries about the inflationary nature of fiat currencies and the inherent insecurity of physical cash.

Self-Interested, Self-Policing Communities

Mining is a built-in quality control and policing mechanism for cryptocurrencies. Because they're paid for their efforts, miners have a financial stake in keeping accurate, up-to-date transaction records – thereby securing the integrity of the system and the value of the currency.

Robust Privacy Protections

Privacy and anonymity were chief concerns for early cryptocurrency proponents, and remain so today. Many cryptocurrency users employ pseudonyms unconnected to any information, accounts, or stored data that could identify them. Though it's possible for sophisticated community members to deduce users' identities, newer cryptocurrencies (post-Bitcoin) have additional protections that make it much more difficult.

Harder for Governments to Exact Financial Retribution

When citizens in repressive countries run afoul of their governments, said governments can easily freeze or seize their domestic bank accounts, or reverse transactions made in local currency. That's not possible with cryptocurrencies, whose decentralized nature – funds and transaction records are stored in numerous locations around the world – effectively prevents state seizure. It's a bit of an oversimplification, but using cryptocurrency is like having access to a theoretically unlimited number of offshore bank accounts.

Generally Cheaper Than Traditional Electronic Transactions

The concepts of block keys, private keys, and wallets effectively solve the double-spending problem, ensuring that new cryptocurrencies aren't abused by tech-savvy crooks capable of duplicating digital funds. Cryptocurrencies' security features also eliminate the need for a third-party payment processor – such as Visa or PayPal – to authenticate and verify every electronic financial transaction.

In turn, this eliminates the need for mandatory transaction fees to support those payment processors' work – since miners, the cryptocurrency equivalent of payment processors, earn new currency units for their work in addition to optional transaction fees. Cryptocurrency transaction fees are generally less than 1% of the transaction value, versus 1.5% to 3% for credit card payment processors and PayPal.

Fewer Barriers and Costs to International Transactions

Cryptocurrencies don't treat international transactions any differently than domestic transactions. Transactions are either free or come with a nominal transaction fee, no matter where the sender and recipient are located. This is a huge advantage relative to international transactions involving fiat currency, which almost always have some special fees that don't apply to domestic transactions – such as international credit card or ATM fees. And direct international money transfers can be very expensive, with fees sometimes exceeding 10% or 15% of the transferred amount.

Chapter 11: Can Ethereum Become The New Bitcoin?

Why cryptocurrency is interesting is difficult to say in one or two sentences, but the fact is that crypts will play a major role in our lives in the future. The issue is when we will be able to pay for services with virtual money (crypts). when we all trade on stock exchanges with cryptocurrency and when it will be equal to money. Do not be fooled - this will happen sooner or later. Already now Bitcoin and Ethereum are trying to rank as currencies in many markets, and this will happen. When this happens, there will be a wave of other crypts that will also be equal to money and internationally recognized.

We believe that this will happen in Croatia, although the law does not yet recognize Bitcoin and other currencies as any currency or value. There is not even a tax on trading with them, which is a very often asked question for the Tax Administration.

Really, recognition of the crypts will first occur in Japan, China, and America, then in Europe and therefore in Croatia. But this text will not be about it, but one of the 700+ cryptos that exists on the market, which is Ethereum - a crypt with great plans, and even more potential. Many think Bitcoin is the only real currency worth mentioning, but it's far away from the truth. Etherum grabs the throne with great strides, and the question is when will reach and pass Bitcoin.

It is necessary to distinguish Ethereum, that is, Ether as a crypt (virtual coin) and Etherum platform on which Ether lives. Ethereum is a public, open source, blockchain platform that works with smart contracts or smart contracts, but before we go into details, we will try to explain in simple terms how Ethereum works. When you use something (for example, a web service, a social network ...) and save your data, they are saved to computers, or servers owned by Google, Facebook, Amazon or some other neighborhoods. Touch companies have a lot of staff specialists who care about these servers, security, data replication, and so on. But this is a server-client model that is very vulnerable.

When the server fails, or it is unavailable, nothing else works. You cannot connect to Dropbox and get to your data; you can not visit Facebook and send a message to a friend, you cannot read emails on Gmail ... Each company has practically a single point of failure or one critical point.

As the creator of Apache Web Server said - this is the original and the first sin of the internet. It is this model that is practically difficult today with all the hackers and attacks that are being carried out. There should be some decentralization of data and servers, and Ethereum is on that track.

On this network, coins called "Ether" are generated, so that nodes can compensate for the use of their resources. If you are a node on the Ethereum network, you receive for all the transcriptions of writing into blockchain a certain portion of the Ether. It is precisely the ether of the crypt that went dangerously to the top, which value drastically increases and which could "download" Bitcoin from the top of the most popular crypts and exceed its value. Today, the Ether one is worth over $ 280, although its value varies. However, given the fact that last year's value was $ 0.20, there is a drastic rise. Not to mention that those who invested smartly and on the stock exchange crypts bought Ethereums are relatively rich today. It depends how much they bought.

If you invested $100 in 500 ETH in the eighth last year, today $ 100 would cost about $ 140,000. Yes, you read well. The same thing is with Bitcoin. The one who invested in BTC 5-6 years ago today can have millions because of 1 BTC = 2800+ dollars. But we missed the opportunity, at least most of us, but also today to invest money in Ethereum; it's a great chance that you will earn a 10- even in the next year or two. All the analyzes say that by the end of 2018, Ethereum should arrive and overtake Bitcoin, which means that the value of Ethereum could be about 3000+ dollars. Let's just mention that every node on the network takes over certain rules (smart contracts) according to which because of. When a particular action triggers a smart deal, it is turned off. If everything is OK and by default, the node in the blockchain (distributed database) writes data. And so, all nodes are doing.

And you have a completely decentralized network where every node knows everything and has a whole database with all the data. The only way that someone can "tear down" the network is to take control of over 50% of the Ethereum network, which is practically and theoretically impossible. All in all, a very interesting concept that will definitely change the internet we know. This is not an invitation to invest money that you do not have, but if you have money that you can lose, your investment in Ethereum could be worth it. With an investment of $1000, you will probably have $10,000 quickly. Of course, the risk is solely on you, because also Etherum can fall to $ 0 and everything invested in it becomes worthless. So invest smartly.

A good question is where Ethereum goes? The platform is two years old, but in those two years, they have done a lot. But it will do more in the future. You can see from the text that they are working on security, cryptography, better and more flexible smart contracts, proof-of-work, mining, new applications, new algorithms, and so on. With each new feature, the value of the Ether will grow, but Ethereum will also grow in a way that many projects will be built on it.

Many countries are considering using Etherum platform for national projects such as bookkeeping and land ownership. Why would not it be written in a distributed database and why it would not be transparent? Then there are logistics projects that could reduce current costs that are measured in billions to just ten million dollars and smaller projects about which we have only information. We also remember Uber. Why would not driving and all money transactions be recorded in a blockchain, and all drivers were knots that would write it down? Indeed, only the sky is the limit for this technology. Everything you think, you can do this like that. You can also look at Sisco's project. It's a distributed and encrypted Dropbox that will be decentralized and could be very interesting in the future.

Interest in Etherum occurred a few days ago in Russia. The main programmer and the man behind the Etherum are Vitalik Buterin, met with one of the most powerful people in the world - Vladimir Putin. Although there is not much information they discussed it, it is evident that Putin was interested in the possibility of implementing Ethereum in Russia and encouraging certain projects. Oil and gas bring a lot of money to Russia, but they also realized that they have to expand the scope of action and that the digital economy the basis for all future tasks to be dealt with. It is also interesting that in Russia, the first pilot project for the Etherum-based bank has been completed, and Sberbank has also participated in it. It suggests that Russia could advance America and become the first big country to accept the crypt of payment as money and use it for faster and safer business, sending and receiving money, and doing business among business entities. Then there would no longer be any non-payment, fraud, and the like, or there would be less of them. In any case, the story about Ethereum is gossiping and will be interesting to follow. Especially through this and the next year when there could be a huge expansion of Ethereum but also other currencies.

Advangates

Integrity (unchangeable) - the third party can't make any changes to the data

Consensus - applications are made on the principles of consensus, and it is impossible to make censorship, and decisions are made unanimously

Security - without a central node, which is usually a bottleneck and if it is unavailable, the whole application becomes unavailable, the applications are ultra safe and can't be hacked

They are always active - the application can never become unavailable or extinguished, no one can shut it down, and even the government

Of course, not all are perfect and with the various advantages of decentralized applications are not perfect. The smart contract is the source code written by the developers, and they are as good as the good and the programmer who wrote them. Errors in the source code made by the developer can cause actions that are executed, but they were not planned and thus not imagined. If there is a failure in an application that is later abused, there is no effective way to correct the default. The only way is that according to the principle of consensus, all involved in the network agree, and the source code is corrected. This directly contradicts the integrity that is ensured by the blockchain, and any such action violates the nature and basic principles of decentralization.

The value of Ethereum has experienced drastic growth over the past year, everyone is talking about Bitcoin, but

Ethereum is behind him. Last year, you could buy Ethereum for $ 10, but now you need to invest more than $ 1000 for that. Although Bitcoin has now become more expensive for many, Ethereum has the potential to break through $ 2,000 and is currently still relatively affordable. Decentralized applications are made from a source code that runs and executes on a blockchain network; it is not possible to control its behavior over a centralized entity.

Any centralized application or service can be decentralized with Ethereum. Credit withdrawal by the bank, registration of users in some kind of register, online voting.

If you want to buy Bitcoin crypts from Serbia, the shortest possible description of the purchase scenario is:

• Choice of exchange for crypts
• Deposit of funds (money in euros or dollars) by payment by credit card or Paypal
• Exchange of funds for Ethereum

- Switching crypts into a safer place (wallet)

This was a brief description of the purchase. And now we are going to a detailed description of each of the steps in shopping and what is important to pay attention!

There are a large number of exchange offices through which you can conduct the purchase and exchange of crypts. All exchange offices can be divided into:

Cash Exchange - This group includes ATMs and merchants.

They installed ATMs through which you can buy and sell crypts. The Mana ATM is that there are few at the moment and it often happens that Bitcoin is not available for sale.

- On the other hand, there are sites where you can personally reach out to retailers from Serbia, the city you live in, which will sell you to the cryptocurrency in return for cash.

Internet Exchange Offices - This group includes online sites through which you can purchase crypts by paying funds through a credit card or Paypal.

Here are some of the factors that you need to pay attention when selecting a currency exchange and purchasing them:

1. Privacy. If you need privacy or do not want to disclose information about you when shopping, then you can forget about online sites and shop through a card or Paypal. Namely, all these sites require a verification process of your identity before purchasing, which includes providing them with an insight into all of your personal information, including scanned personal documents. This has become a practice, and it is not necessary to be afraid of this because of they all function like that, but it may simply be difficult for someone to do this because the process is slow, so it's good to keep in mind. In this case, the best option is cash exchange, that is, buying through ATMs or personally through traders.

2. Limitations. This only applies to those who want to buy a larger amount of crypts

at once. Namely, all exchange offices have certain limits that allow only certain quantities of crypts for purchase. Since we have less payment power from Serbia, this factor is not likely to pose any problem to the purchase.

3. Speed. The process of buying cryptocurrency can be slow and not so easy. Namely, as I have already written, online exchanges will require you to verify the identity first before you buy, and as this can take a lot of people who want to make purchases in the past, this may take several days. On the other hand, shopping via ATMs or through retailers will be the fastest, as soon as money is transferred, crypts are sent to your wallet.

4. Exchange course. There is no official and fixed course of crypts. It depends on the demand and the offer depending on the exchange office to the exchange office. The most relevant site where you can track the average value of all crypts is CoinMarketCap. Here too, you can see for each crypt on which exchange is best to do shopping (where is the best course).

5. Scams. You need to be very cautious when shopping. Like everywhere, there are a number of fraud options. For this reason, it is important that you choose the most trusted exchange offices, which is easiest to determine by the number of transactions totaled by an exchange office so far. However, I must mention here that it is very important that you do not keep your crypts on longer paths within the exchange offices. It would be best to transfer them to your wallet after purchasing.

6. Commissions. Each exchange office has its own commission, and it is very important to check exactly what commission is before the purchase because it can be significantly high. The commission generally exists for every step of the purchase, that is: when you pay your funds when exchanging funds for crypts, as well as when transferring crypts to your wallet.

The crypto wallets contain private keys or security codes that allow you to manipulate the wallet and money in your wallet. Each wallet generates your public address that is related to a specific cryptocurrency. This address is unique, and you will use this address when you want to receive from someone steam (for example, when you buy crypts in currency exchange, the currency exchange will send you the currency via that address). To access your wallet, you will generate a security code, which you will only know. You can save it on your computer, in a file, or print it on paper.

There are several types of the digital wallet.

Physical (Hardware) wallets - they are a small device like USB sticks. Their purpose is to serve as your money for crypts. To work, they need to be connected to a computer, usually via a USB port. The most popular are: Ledger Nano S, KeepKey, and Treasury. I will write more about them in a separate article. What matters to this type of wallet is that they are a good choice when it comes to security. They are easy to use. The only disadvantage is that you need to buy them. It is recommended to store larger values of crypts in a wallet. Security codes are stored within the wallet itself, so only the owner can have access to the wallet if he/she has a password to unlock the security code.

Internet banknotes - they are banknotes applications that work on computers, phones, or tablets. They are very easy to use and are recommended when you want to save fewer crypto values in your wallet. The security code is stored on the device through which the wallet works, so this wallet is not considered 100% secure.

If you invest a large amount of money in the purchase of crypts and want to keep them in the long run, then physical wallets are the right choice for you. Otherwise, if you want to trade with crypts, then the online wallet will satisfy your needs.

Here I must point out that you must make the difference between the internet wallet and the internet exchange. Within both, it is possible that you have your own crypts, but they are much safer in your banknotes. Exchange offices are risky because no one can guarantee that they will be extinguished for some reason or for some reason. And so your crypts will disappear.

Another note. At this moment, there are more than 1000 available crypts on the market, next to Bitcoin and Ethereum, which are the most popular. What is important is that each wallet does not provide the ability to store each crypt. This way, depending on your wallet, you will be able to keep only one or more specific crypts in one.

Chapter 12: How To Get Ethereum Coins

So now you've read in previous chapters what cryptocurrencies are, what Ethereum is, learned a bit of basic history about them both, as well as come to understand what it is they do exactly. Assuming that you're continuing to read this chapter and show interest in investing and trading in Ethereum, the next logical step would be to discover how to obtain Ethereum then

After all, it's not like any other form of investment or currency. You don't get a hold of it, and physically hold it in your hand to use whenever you want. It's also not an investment option that has any form of certification like stocks, bonds, or IRA's or 401(k)'s either. It's all held digitally, and that means that you own it on your computer.

Fair enough. But how do you even start to obtain them? It has to come from somewhere. Physical currencies are simply IOU's printed off from the government based on the strength of its economy. You can't print off or copy, or digitally get a hold of Ethereum coins through normal means.

Luckily, though, they're three major ways that you can get a hold of Ethereum coins. We'll be taking a quick look at each of their applications, how to earn them, and even a little bit of information about their pros and cons.

Mining

Technically, Mining isn't what you think it would be. It's not about grabbing heavy tools, and going out and digging in the earth for the valuable Ethereum coins and all. As has been mentioned before, Ethereum coins are an entirely digital thing that you can only get through the internet, and can only hold on your computer.

So then what exactly IS mining for coins exactly?

Mining is a technical term. In normal circumstances with physical currency, governments print more money and base it off the strength of their economy. With cryptocurrencies though mining is one of the main ways that they're able to make/discover more altcoins. Of course, like with traditional real life mining you will be competing with other miners to earn coins.

How does it work?

Simple. Whenever we send Ethereum coins to one another or do any sort of transaction with the coins, a little bit of information known as a hash gets left behind. Collecting enough of these hashes on the end of each transaction eventually gets collected by the Ethereum network and stores itself as a sort of "block", which miners then use to chip away at using mining applications from their computer to earn bitcoins a little bit at a time before eventually the entire block falls apart, the parts are distributed, and everyone gets their share that they put in.

However, keep in mind that the basics of that are a lot more complicated than simply finding the block and having access to it. That's more or less a simplified version of how the process goes, and mining is very CPU intensive. You essentially need networked computers put together, or at least a CPU that's built almost exclusively towards mining the Ethereum coins. Luckily, though, there's the possibility of joining a mining pool where everyone pools resources together and shares in earning coins together.

All in all, though, while Mining is very complicated (and deserves it's own e-book in its own right), it's also very slow and takes an initial investment just to get started. While you can literally start with nothing but a powerful PC and internet access, there are other ways to invest in Ethereum.

Trading

Trading is a very lucrative way of earning Ethereum coins. Simply put, you have a goods or service that someone wants, and they give you a payment for that. It's really that easy. It's just like traditional currency transactions.

The only downside to it is that Ethereum coins fluctuate in prices on a daily basis. One day you'll be earning more coins due to the fluctuation of the market, while the next day you'll earn considerably less than what you intended before the market goes on an upswing. This happens normally in real life daily too, as most currencies are in a state of fluctuation, but there are usually limits in place to keep it from being too drastic too quickly.

Let's put this into an example. Let's say you offer I.T work for professionals and help maintain networks. You charge 5 Ethereum Coins at the current market value of what they're worth. They go up in value, meaning that the Ethereum you earned before are worth more than they were a bit ago. At the same time though, while the Ethereum coin is worth more, your services are still judged based on outside influences, such as other real world currencies (Such as the U.S Dollar, or EU Euro) and so, your services are expected to remain the same price, thus lowering your prices from 5 Ethereum to a possible 3 Ethereum.

Technically, that's not bad. You're still earning the same as you were before, just less Ethereum coins. But, don't be so hasty to celebrate your good fortune, because soon the market swings against your favor and Ethereum coins drop below in price, well below what they initially started at when you offered your I.T services. Now the three that you earned from before are worth considerably less than before, meaning you're earning less for the same amount of work.

Of course, the upside of this is that you can again adjust your price to even higher than you had them at than initially, and the market might once again swing in your favor, making the coins you saved worth even more than before.

Which brings us to another way of getting Ethereum coins that are similar in the same vein, but entirely different.

Investing

Investing in Ethereum coins is pretty simple. You simply purchase them based on their current monetary value, trading them out for another currency, and then you hold onto it hoping the value of Ethereum coins rise's enough to have justified the purchase. That's it. That's all there really is when it comes to Ethereum investing.

Though, it does get a little bit deeper than that to be quite honest. It's more than just holding onto them and hoarding them. After all, if everyone just hoarded their Ethereum coins like some sort of digital dragon, then they wouldn't really rise in price after all. They'd just be expensively and needlessly complicated toys. With investing you have to at least be a part of the reason to stimulate the market. So generally it's acceptable to take your invested Ethereum coin, and even use it as part of the market itself.

However, at the same time, it's through investing such as this that drives the price of Ethereum coins up, and makes it more valuable as well as their use as an actual currency that draws people in. So don't be afraid to take what you've invested in, and even use some occasionally to stimulate the market for other goods and services, or even sell some off yourself to other interested parties should the value increase. The more people that are involved, and the more people are invested in the Ethereum platform, the more they'll be worth in the future.

Chapter 13: Understanding Gas

In this chapter, you are going to learn about what gas is and how it works.

The Ethereum network brought with it several new concepts that do not exist on other cryptocurrency networks. One of these is something known as gas. This is something that most people have a hard time understanding. Whenever you want to make a transaction on the Ethereum network, you will notice that you will be prompted to enter a gas price and gas limit (some wallets automatically decide on the gas price and gas limit to keep things simpler for users).

So, what it this gas, and how does it work? Whenever you perform an operation on the Ethereum network, whether it is sending Ether or interacting with a smart contract, some computations will have to be performed by miners on the Ethereum network for your operation to be successful. For their work, the miners need to be paid. This is where gas comes in. Gas is a unit that is used to measure the amount of work (computation) that had to be done for your transaction to go through. However, the payment for the transaction is paid in Ether.

To make this concept easier to understand, let us compare it to using electricity in your house. Whenever you wash your clothes in the washing machine, the machine uses electricity, which you need to pay for. You can think of the electricity as the computational resources needed for Ethereum operations. However, the electricity used up as you do your laundry is not measured in dollars. It is measured in Kilowatts per hour. Similarly, the amount of computational resources needed in order for your Ethereum transaction to go through are measured in gas, instead of Ether. Finally, despite being measured in kilowatts per hour, you pay for your electricity consumption in dollars. The same way, despite being measured in gas, you pay for the amount of computation done for your operation to go through in Ether, not gas.

Every operation on the Ethereum network requires gas. However, different operations require different amounts of gas, depending on the amount of computation that must be done for the computation to go through. Simple operations like sending Ether need only some little gas, while more complex operations like interacting with multiple smart contracts need higher amounts of gas.

You might be wondering why the payment for transactions is measured in terms of gas instead of simply charging a certain amount of Ether per operation. Ether is a token that is publicly traded on cryptocurrency exchanges, therefore the value of Ether is subject to rapid market fluctuations. At the same time, the amount of computational resources required for transactions does not change as rapidly. This means that a 20x increase in the price of Ether would lead to a 20x increase in the amount paid for the same transaction. This is not very sustainable. The gas system was therefore created as a solution to keep the computation fees independent from the price of Ether.

Gas Limit

Before performing any operation on the Ethereum network, you will be required to set the gas limit for your transaction. This determines the highest amount of gas you are willing to spend on the operation. Gas limit is a precautionary mechanism that keeps you from using up all your Ether in a faulty operation. For instance, if you execute a smart contract that has a bug, it would keep running without achieving the desired objective. However, since it is still running, it still requires computational resources, so you would keep paying for an operation that is not going anywhere. However, by setting a gas limit, the faulty operation will only continue until runs out of gas, at which point it will stop.

The recommended gas limit for standard transactions is 21000. If you set the gas limit for your transaction as 50000 and your transaction only uses up only 21000, all the extra gas will be refunded to you once the transaction is complete. Even though it is up to you to set your preferred gas limit, you cannot lower your transaction cost by lowering your gas limit. The amount of gas required for a transaction depends on the amount of code that must be executed in order for the transaction to go through. Therefore, it is not up to you to determine how much gas the transaction will use up. If you set too low a gas limit, the transaction will run out of gas without being completed. The used-up gas won't be refunded to you. Instead, it will be kept by the miner, since they have already spent their computational resources executing your operation.

Gas Price

Apart from setting the gas limit, Ethereum transactions will also require you to set the amount you are willing to spend for gas. Gas price is usually measured in Gwei, which is a fraction of an Ether to the 9th decimal place. The ability to determine your gas price means that you can lower your transaction cost by entering a lower gas price. However, the system is made in such a way that it encourages people to use a reasonable gas price. The gas price you decide to pay for your transaction determines how fast your transaction will be mined. How does this work?

For every transaction on the Ethereum transaction, the transaction fees are paid to the miner who provide the computational resources for the transaction. Whenever you initiate a transaction, it is entered into a pool from which miners then pick transactions to include in a block. It is up to the miner to decide which transactions to include in the block. As such, it is only logical that they will first choose the transactions with a high gas price since they get the highest returns from processing these transactions. Therefore, if your transaction has a high gas price, the high gas price will be an incentive for miners to process your transactions before those with a low gas price. If you set a gas price that is too low, no miner will be willing to add it to a block, so it will remain as a pending transaction. This means that you should set a gas price that is high enough for miners to want to include it in a block. If you need your transaction to be processed faster, you can set a higher gas price. The actual transaction cost for a transaction (paid in Ether) is a factor of the number of units of gas used for the transaction and the transaction's gas price.

Chapter Summary

In this chapter, you have learned:

Gas is a unit that is used to measure the amount of computation that had to be done in order for your transaction to go through.

Though transaction costs are calculated in terms of gas, the payment for the transaction is paid in Ether.

Different operations on the Ethereum blockchain require different amounts of gas, depending on the amount of computation that must be done for the computation to go through.

The gas system was created as a solution to keep the computation fees independent from the price of Ether.

The gas limit determines the highest amount of gas you are willing to spend on an operation on the Ethereum blockchain.

Gas limit is a precautionary mechanism that keeps you from using up all your Ether in a faulty operation.

Gas price is the amount of Ether you are willing to for gas.

The gas price you decide to pay for your transaction determines how fast your transaction will be mined.

The actual cost for a transaction on the Ethereum blockchain is a factor of the number of units of gas used for the transaction and the transaction's gas price. In the next chapter, you will learn how to buy Ether.

Chapter 14: How To Invest In Ethereum (And Is It Too Late)

Ethereum has become a popular cryptocurrency alternative to Bitcoin over the last year. However, unlike Bitcoin and rival currency Litecoin, Ethereum has been adopted by many companies and startups as a way to transact (and more).

In the cryptocurrency wars, I like to view Ethereum like the diamond of the currencies - it has both a intrinsic value and an industrial value.? Compare this to Bitcoin, which operates like gold - not much industrial value, but people buy it and sell it based on it's intrinsic value to the holder.

Given the popularity of Ethereum, many people are curious about what it actually is, how it's different than Bitcoin, and how to invest in it. It's also important to note the risks of investing, and the potential to mine it and create your own wealth of Ether (the actual monetary unit of Ethereum).

Ethereum is basically software that is decentralized and allows ?developers and programmers to run the code of any application. Wait, what? I thought Ethereum was money... well it has a monetary aspect.

You see, Bitcoin uses a technology called blockchain specifically for conducting monetary transaction - it's a straight currency. Ethereum uses blockchain technology to allow the creation of applications that can be executed in the cloud, can be protected from manipulation, and much more (some stuff getting too technical for me here). However, a bi-product?of this is that Ethereum uses a token called Ether, which is like Bitcoin, to transact. This is the monetary value portion of Ethereum.

Because of it's unique abilities, Ethereum has attracted all types of attention - from finance, to real estate, to investors, software developers, hardware manufacturers, and more?

Ripple is similar to Ethereum in that it's token XRP is also able to conduct real transactions.

If you're interested in investing in Ethereum, and specifically Ether, you need a digital wallet. Ethereum doesn't trade on any major stock platform. You can't go to your online discount broker and buy Ethereum. You have to convert it into your wallet.

We recommend using Coinbase as a digital wallet because it's incredibly easy to use, allows you to invest in Bitcoin and Litecoin as well, and they will give you a bonus for signing up. If you sign up with this link you'll get a $10 in Bitcoin bonus if you deposit $100.?

It's important to remember that Ether (ETH) is a currency, and should be treated as such by investors. You don't buy shares of Ether like you would stocks or ETFs. Instead, you are exchanging your dollars for Ether tokens. There are no dividends, no payouts. Your only hope is that in the future, other people on the Internet will pay you more for your tokens than you bought them for.?

Final Thoughts

Investing in Ethereum is risky, but it could potentially be lucrative. Unlike Bitcoin or Litecoin, companies are really using Ethereum as a building block - something more akin to diamonds than gold. As an investor, this is a potential win.

Furthermore, there can be splits (i.e. hard forks) on Ethereum like we recently saw with Bitcoin and Bitcoin Cash. This can be a good thing or bad thing. People who've invested in Bitcoin Cash are happy about the split because they made great money for no effort.

Why is Ethereum valuable?

Unlike other assets, Ethereum is not backed by gold or promised by government. To understand whether Ethereum is worth buying, it is first best to examine the fundamental value of the Ethereum blockchain itself. For the sake of simplicity, this section will look at the Ethereum blockchain only.

Mathematics and scarcity

The Ethereum blockchain is a protocol that operates on the laws of mathematics. Unlike a central bank or government, who can quickly and unexpectedly adjust money supply, Ethereum's coin distribution is written into immutable code that is publicly available and agreed by consensus. It is the blockchain's unbreakable encryption and mathematical truths which back this digital asset, as opposed to gold or government promise.

Ethereum is an inflationary currency; 5 new Ether coins enter the system whenever the next valid block in the blockchain is found (a block is found roughly every 15 seconds). The process of finding blocks is a separate topic, but the key point is that – unlike Bitcoin, whose supply is capped at 21 million coins – there is no limit on the amount of Ether that will be issued over time. However, this rate of inflation will decrease over time as the aforementioned issuance of 5 Ether becomes a smaller percentage of the overall coin supply. Furthermore, planned network changes (which must be agreed by consensus), due to be launched in the coming months, will place downward pressure on the inflation rate as discussed here.

Sovereignty

Transactions on the Ethereum blockchain are valid based on a few factors, but the most obvious is that the user must have a balance greater than the amount they are sending. The purpose for which they are sending or receiving coins is irrelevant. Any user of the Ethereum blockchain – regardless of location – is able to decide how to spend their value without authorization. Having sovereignty over one's wealth may seem unnecessary for many in the West, however those from developing nations, or countries experiencing hyper inflation and money controls, stand to benefit enormously by untethering from their fiat currency system. Unlike the traditional fiat system, Ethereum offers users full sovereignty if they wish. Of course users can choose to trust 3rd parties if they would like to, but that is not a requirement as it is in the traditional banking space today.

Efficiency

Ethereum transactions are low cost and fast, capable of handling 15 transactions per second with protocol upgrades in the next 12 months that are anticipated to increase this figure to 1000+. To put that into perspective, VISA handles an estimated average of 2,000 transactions per second. Furthermore, 3rd party payment channels are being developed which will take transactions off of the Ethereum blockchain without compromising security and reducing fees further – increasing the capacity of the network by several orders of magnitude.

Liquidity

Ether has real-world value that is in demand. Major Ethereum exchanges will complete large million dollar sell orders within seconds without moving the price. Liquidity could certainly be higher, and brief "flash crashes" have been noted in the past, however for the vast majority of users and investors, Ethereum's liquidity allows for fast exchange to and from fiat currency.

Ethereum Virtual Machine (EVM)

Up until now this article has focused on the fundamentals of the Ethereum blockchain and its use case as a currency for transacting value. Ether serves well as a currency, however it is the ability to deploy "smart contracts" on the EVM which furthers its case as an alternative investment. Smart contracts are still in their infancy, however a number of industries are on the cusp of major disruption thanks to this technology:

Prediction markets

Gambling

Insurance

Trading

Much like gold, Ethereum and others are being used as a hedge against economic uncertainty. However unlike gold, Ether can also be transacted globally and near-instantaneously through the internet with minimal fees and unlimited amounts. The supply of Ethereum is also transparent and predictable through its open source code which is publicly auditable. In the case of gold, supply shocks are not uncommon.

Selling Ethereum

Many investors choose to buy and hold Ethereum for the long term, predicting that the price will rise in the years ahead. However, at some point in the future it may be prudent to sell some or all of an ETH portfolio. Here's how it works.

Chapter 15: Ether And Other Crypto-Currencies

Crypto-currencies have been well explained by their founders and computer scientists, yet they are still a mystery to the rest of the general public. Those who have studied crypto-currencies are well aware of the importance they will have in the future. In this chapter, we will look at what crypto-currencies are and the market in which they are traded.

Crypto-currencies are digital money that is transferred without the help of a central bank. The currencies are encrypted to prevent fraud or control from a central point. We can take crypto-currencies as entries in a ledger that cannot be changed once a transaction is complete.

Some of the examples of crypto-currencies include, but are not limited to, Bitcoin, Litecoin, and Ether. These are just a few of the crypto-currencies available on the crypto-currency trading market. The crypto-currency market is a platform for the buying and selling of digital currencies all over the world. It can be compared to the foreign exchange market, where currencies from different countries are traded. The difference is that the digital currencies are not affiliated with any country, region, organization or location. The currencies are fully decentralized.

On the crypto-currency market, you can trade your Bitcoin or Ether for USD or BTC. This way, you can be involved in the cryptocurrency world without the risks that come with mining. Cloud mining and other such things will be discussed in detail further in this chapter.

Just like the standard foreign exchanges, there are different sites on which you can trade for your crypto-currencies. Some of them are more complex than others. These are the ones that are used by professionals to trade cryptocurrencies. They give access to all the fancy tools for trading. If you are not interested in having the tools, then there are sites where you can trade without the need for creating an account.

Once you have decided what kind of trading you want to do, then the next step needed is a decision on the type of exchange you are going to use.

There are three types of exchanges.

Direct Trading—This is a platform where one person exchanges currencies directly—from one person to another, often in foreign countries. There are no fixed foreign exchange rates, and the participants of the trade set their own.

Trading Platforms—These are platforms where buyers and sellers are connected to each other. The platform often takes a fee for every transaction.

Brokers—You can visit these sites and buy the crypto-currencies directly from the dealer. The price is set by the dealer. We can compare them to the regular foreign exchange traders.

Before you join any exchange on the internet, you need to look out for a few things:

Country Restrictions

The business of digital currency is for the whole world. The exchanges are, however, not as flexible. Some of the functions offered by some of these exchanges are only accessible by being located within a particular country. This is important, especially if you are planning to buy any currencies. Check that all the services of the site are available in the location that you are currently at.

Exchange Rate

This one may seem rather obvious but is crucial to trading. Some platforms change their exchange rates regularly. Check for such things before you decide to join any platform.

User Fees

As we have mentioned before, make sure that you understand the fees taken from your transactions. For some of the platforms, the fees for depositing currency are minimal but get higher when you withdraw. Check on such things before joining any platform.

Payment Options

The more payment options available, the better for you. If you find one with limitations on the number of payment options, then you need to look for a different platform. The platforms that only offer payments from debit or credit cards can be susceptible to fraud. You would need to give out your credit card details, and that could be risky.

The Methods of Verification

Some of the exchanges will require you to verify your user ID. Some of them will allow you to remain anonymous. The process of verification may take longer, but is worth it. The time required to verify your identity before depositing or withdrawing is a way to prevent fraud attempts.

There are various advantages of trading on the crypto-currency market compared to the traditional foreign exchange.

The spreads are smaller

On the regular market, the transactions you make alone could eat into your money due to the spreads. Spread is the difference between the asking price and the bid price of the currency. Even when you use Euros or US dollars, the spread on your transaction can be large. With cryptocurrencies, the range is minimal, and it's almost as if there are no fees for the transaction.

Margin Trading

You are allowed to trade on margin in the crypto-currency world. What that means is that you can pay some amount of money and get the rest of your funding from your peers. The money lent is returned to the lender with some form of interest but allows you to make a profit with money that is not yours.

Leverage Trading

This is trading with money that you currently do not have. The average rates for leverage trading on most crypto-currency sites are 1:10. This means that, if you have 1 dollar, you can trade it as 10 dollars even though you do not have that amount.

Chapter 16: Dao And Ethereum Classic

DAO attack

Now that you have started learning about Ethereum, you need to know about a very crucial part of the creation of the decentralized platform. As mentioned earlier, there was an attack on one of the crowdfunding organizations on the platform, known as the DAO. DAO is an acronym for Decentralized Autonomous Organization. From the name, it is clear that the organization has no particular central power or owner. The decisions on how to invest the money that is crowdsourced are left to a vote. The amount of contribution gives more power to the voting party.

The purpose of the organization is to have a body that replaces governing agencies in a decentralized way. The tokens that the individuals purchase to become part of the DAO are not equity shares that reflect ownership. The tokens are merely contributions that allow you to have voting rights.

The DAO was an organization created to fund the development of certain applications on Ethereum. The founders of a Smartlock company started this particular DAO and named it The DAO.

The DAO began with high expectations to get funding that would then be used to develop several apps. They, however, did not foresee the mostly positive response to their crowdfunding. For some reason, the Smartlock company founders attracted a lot of investors. Some people claim that because they did not expect such a significant investment they were not well prepared. The unpreparedness for such a response is one of the reasons why the attack happened. The crowdfunding was one of the fastest and largest in history. The founders managed to collect over $150 million!

The people who were part of The DAO aimed to pool their resources together to invest in projects on the platform that were set to make a profit. The

Decentralized Autonomous Organization was intended to be a transparent team where any of the members could audit and see the code. The expectations of the DAO and its performance, in reality, however, played out very differently. With such a significant amount of money invested in their crowdsale, various problems arose that the founders did not easily see until the attack had already happened.

The group purchase was made in the form of a smart contract. Individual parameters were set to be fulfilled. On June 9th, one of the developers on the network pointed out a few vulnerabilities in the DAO, and this was blogged by the founder of the blockchain network. Other vulnerabilities were also noticed, and proposals to fix the problems were ready by the June 14th. These proposals were awaiting the approval of members from the DAO. On the 17th of June 2016, a hacker managed to use some of the loop-holes in the smart contract to siphon $3.6million of the investors' Ether. The hacker is said to have voluntarily stopped the process once people figured out what had happened. The money was siphoned into a child DAO. As the smart contract stated on the Ethereum network, the money was held for 28 days. The money was, therefore, not lost. The money is still in the child DAO as any movement of the crypto-currency will trigger investigations. The Ethereum founders and some members of the DAO decided to 'hard-fork' the blockchain. This was because the attack on the network caused the value of Ether on the market to drop from a rising $20 to below $13. This was a big blow to the quick development of Ethereum as a platform. Its integrity, which was untarnished before, was suddenly in trouble and started losing value.

The members of the DAO had to come up with a solution as to whether to hard fork it or recall all the money back to the original address. A hard fork meant secluding the DAO from the rest of the network. The crypto-currency on the network was now known as Ether (ETH), while the Ether that was part of the DAO attack was now known as Ethereum Classic (ETC). You can now buy both of these on crypto-currency exchanges.

Ethereum Classic was a result of the people who disagreed on the hard-forking of the blockchain, saying that the blockchain could not be changed. The Ethereum Classic chain was then hard-forked internally to adjust some of the pricing of the operational codes on the network. Various changes have been made to the system after this hard fork and more are scheduled to come. The Ethereum members who rejected the hard fork now have ETC while the rest of the network has ETH. The two have different values on the market, but Ether is traded as a combination of the two.

Ethereum Classic also made changes to the way the blockchain worked. They did another hard fork to eliminate the 'difficulty bomb' on the network. This was the code that made mining of Ether difficult.

Chapter 17: Buying And Using Ethereum

In this chapter, we're going to be discussing the different ways in which you can buy and trade Ethereum, as well as the best practices for doing so.

The first thing that you're going to need before you do anything else is a wallet. No, I don't necessarily mean a wallet that you keep in your pocket; cryptocurrencies are kept in things which are called wallets. These can take one of two forms. The one you'll want to get will vary heavily depending upon what exactly you want to do.

The two forms of wallets are soft wallets and hard wallets.

Soft wallets are simply software wallets. These are wallets which may take one of two forms: they may be either an independent piece of software on your computer which may be accessed offline or they may be based on an online service such as an exchange or something of the like.

Soft wallets are not bad, especially if you're not really intended to have any sort of serious amount of money within it. If you're just looking to do small transactions, then a soft wallet will be perfectly fine. However, I will say that I do highly recommend getting a software wallet instead of just leaving it on an exchange wallet.

This is because when you leave your cryptocurrency in the wallet on an exchange, it is at the whim of the exchange until you move it off-site. This doesn't mean necessarily that they'll take it. However, there have been instances where cryptocurrency exchanges have been hacked, and people's wallets have been bled dry because they were compromised whenever the site was.

There are several different Ethereum soft wallets that you may use. Perhaps the best is Exodus. Exodus is a beautiful soft wallet that is extremely easy to use; you won't be asking any questions while you're using it.

It leaves you in complete and total responsibility of your coins. However, this private key to your wallet, the contents also means that if you are to lose the thereof will be lost forever.

Hard wallets are something else entirely. Hard wallets are intended for offline storage of cryptocurrency. They usually attach to your computer by way of USB. When they are disconnected, they are completely and entirely secure. These are the best option if you want to transfer or invest in large amounts of cryptocurrency because it is the option which is least likely to leave you wide open and vulnerable to attacks.

It's hard to say exactly which hard wallet is the best because there are a whole lot of different options in that realm. All of them are extremely easy to use and are pretty much plug-and-play.

So how does one actually buy and use Ethereum? This is where things get a little trickier.

Remember how in the first chapter, we mentioned that cryptocurrency exchanges are pretty much untraceable? This isn't necessarily true. The exchange of cryptocurrency isn't entirely anonymous; rather, it's pseudonymous. This means that every person, instead of having their personal information displayed, instead has a simple address that is designated to their specific wallet. Whenever something is transferred to another address, it is written down in the blockchain and recorded for pretty much all time. (Or as long as that blockchain exists, anyway.)

Every Ethereum wallet has a receiving address. This is the address to which coins may be sent. To send something to another person, you will need their receiving address. If you're using online marketplaces, generally they'll have an independent wallet which you will send coins to that you can then use to purchase things on the site. Regardless, whenever a transfer goes from wallet to another - of any kind - it will happen because somebody sent something to another person or entity's receiving address.

This is the basic mechanism for trading cryptocurrency, Ethereum included. It usually will take place from whatever wallet that you're using, whether it's a soft wallet or a hard wallet.

So at this point, you may be wondering: how does one actually get Ethereum? The simplest way is to do so by using an exchange. Cryptocurrency exchanges are places where you can trade your fiat currency for somebody else's cryptocurrency.

For the purposes of this book, I will specifically recommend Coinbase. All that you do is you sign up for an account which will thereby create a wallet. After that, you just connect your debit card or bank account. After that, you can simply buy cryptocurrencies with extreme ease.

After they're in your exchange wallet, you can transfer them to your offline wallet - whether it's a soft or hard wallet - simply by sending it to the receiving address of your other wallet.

From here, as an investor, your best bet is to either sit on your cryptocurrency or use it to fund Ethereum-related ventures. The way to find the latter is by becoming active in the Ethereum community and seeing what ventures are hot.

Chapter 18: Programing In Ethereum

If you are anxious to get started creating your own application on the Ethereum blockchain then the first thing you are going to want to do is to familiarize yourself with the Solidity programing language which you will find broadly similar to JavaScript. It takes advantage of both the .se and .sol extensions along with LLL, a Lisp byproduct. If you have previously used Serpent or Python, you will feel right at home with Solidity.

In order to ensure that you can compile the apps that you create as easily as possible, you are going to need to choose a solc compiler that works with C++. If you don't like working with solc then you can use an in-browser alternative such as Cosmo instead, though this chapter will assume you went with solc. Additionally, after you have finished compiling your work, you will then need to use the Ethereum Web3.ja API to ensure that you are able to use JavaScript to connect your smart contacts to the app directly. This, in turn, will ensure you are able to interact with your smart contracts wherever you are, not just when you are using an Ethereum node.

Frameworks

When it comes to using a framework for your distributed application, you will likely be relieved to know that there are already a wide variety of options available, for free, that the community has created. These make a terrific place to start and ensure you don't need to worry about building your framework from scratch.

Truffle: Truffle is a great place to start as it automatically completes many of the more generic programming steps required to get a decentralized application up and running. This, in turn, means that you can spend less time going through the motions and more time working on whatever it is about your application that is sure to take the world by storm. If you are using Truffle, then you may also be interested in Embark. The two work together and Embark is known to come in handy when it comes to streamlining and building apps as it automates a majority of the testing process.

Meteor: If you are looking to improve the stack, Meteor is the choice of many Ethereum developers as it works well with the Web3.js API. It also easily works with many standard web application frameworks and was one of the main supporters of the platform during its early days.

API: The API that most new Ethereum developers go with is the one created by BlockApps.net. It largely works the same way as a real Ethereum node, which is perfect for times when you are not in a position to run a real node but still need to do some work on your application. An alternative to this is known as MetaMask which makes it easy to run the standard array of Ethereum tools directly from a web browser. Another option is LightWallet which is an easy way for users as well as developers to interact with decentralized applications while at the same time providing different types of users with different specialized interfaces.

Creating your App

In order to get the application creation process started, the first thing you are going to need to do is download your very own Ethereum node. You can do this directly from your command line by interacting with the Ethereum node interface. On your command line, enter the following: bash<(curlhttps://install-geth.ethereum.org).

With that done you will then be prompted to begin installation after you select the correct operating system and the most up to date version of the Ethereum CLL. Once the installation is complete, you will then be able to access Geth in a JavaScript environment that responds to standard console commands. If you wish to create specialized console commands they will be saved, which will help you track your progress more easily in the long run. With that done, it's time to get to work. You can open the terminal tool to open the Geth console. Once you have done so you should see a > in the corner to indicate things are running smoothly. To quit, simply type EXIT and then hit ENTER.

Once you have finished coding your application or smart contract that will run in your application, the next thing you will need to do is compile it. After everything is complied, you can then deploy the results. To do so, all you need to do is to pay a gas fee and sign a digital contract. Once this has been completed you will be presented with a URL that links directly to the contract's location in the blockchain along with the ABI for the application.

After you have this ABI you will then be able to look at the application or contract from any internet enabled device. Depending on what your application does exactly, each time you interact with it you may be required to pay a fee in gas.

Testing

If you are building a smart contract then it is extremely important that your if/then statements are presented in such a way that there is no wiggle room surrounding them at all. This is a good time to load up Truffle as it will automatically create the type of framework required by both Web3.js and JavaScript to ensure the smart contract works as intended.

When it comes to testing the transaction time, the promises that you use are going to play a large part in whether the smart contract or app actually sees any use by the community. In order to catch on, your application is going to need to verify as quickly as possible. The lower limit is currently 10 seconds, though this is rarely ever seen outside of testing scenarios.

Once you are ready to start testing a contract you will want go to the test directory and change the .js file extension to conference.js, making sure to change any other related references as well. Once this is done you will then want to run Truffle in the root directory that is connected to the test file.

With this completed you will then need to open solc, Solidity and Pip. Prior to doing so, make sure your main library has been separated from the test library through the use of a virtual environment. With this completed, you will then want to open a console window and initiate a new node client.

From there, you are going to start Truffle and use its deploy command to activate the init that will be used with the smart contract. This will make it possible for the program to note errors that the code might contain as of this iteration. While you are developing your code, you are going to want to go ahead and test the compilation in Truffle as well as a means of weeding out any issues before the final compilation. Truffle also tests the deployment of contracts in a virtual space as well.

Deploying

After your contract has been tested successfully, you will also be able to deploy it from Truffle. To do so, begin by going into the console and using the command code truffle init (new directory) which will create a clean directory. With this done you will want to find the contract you have created by looking for the name of contract and the .solc extension. You will then need to find the app.json/config file and add in the name of the contract where it says "Contracts".

You will then be ready to start up your Ethereum node, do this from another window in the console and then use the tesrpc command. Finally, all you need to do is run Truffle one last time and then choose the option that will deploy via the root directory.

Variables to keep in mind

The variables that you use in your smart contract will always organize in the same way. The first variable you come across is going to be the variable for the address which references the location of your ether wallet, assuming it is the primary ether wallet for the contract. This address will automatically generate along with the contract, you can find it by using the () function. Using Geth, you can create different account address from within the same node.

The second variable will always be UINIT which stands for unsigned integer. This variable will almost always be referenced as 256. For most types of smart contracts, you aren't going to have any reason to alter this variable.

The next variable will be listed as either Public or Private which will allow you to decide what your smart contract will be able to access. If it only needs to use information that is going to be found in the blockchain then you can set it to private. If you need your smart contract to outside external details, then you will need to use what is known as an oracle and then set this variable to public.

Chapter 19: Ethereum Smart Contracts

Blockchain was first created to be used with bitcoin because it is the first cryptocurrency. But, when Ethereum came around, there was technology created that would allow for smart contracts to be created. Ethereum is the cryptocurrency that will focus on peer to peer transactions.

In this chapter, you will learn about how to create an Ethereum contract despite the blockchain that you are using because the process will be the same.

You are also going to learn how to analyze you are your contract so that you can ensure that it is written the way that you want it to be before it is placed on the blockchain to be executed.

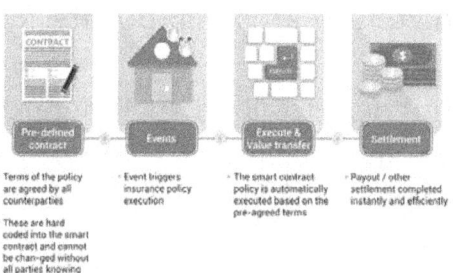

- Terms of the policy are agreed by all counterparties
- These are hard coded into the smart contract and cannot be chan-ged without all parties knowing
- Event triggers insurance policy execution
- The smart contract policy is automatically executed based on the pre-agreed terms
- Payout / other settlement completed instantly and efficiently

Example of smart contract in the insurance field

Definitions

There will be a few terms that you must know to deploy your contract to the blockchain for use.

Public key cryptography: the public key will be made up of a two-part system. The private key is an open key as well. Individuals will have to make a virtual signature that will mark all their blockchain work. It is vital that the users create backups of their keys since they are not going to be able to access their accounts without them. And, there is not going to be any way to access them externally.

Ethereum virtual machine: Smart contracts will be written with the blockchain infrastructure that is appropriate for that contract.

Dapp: this decentralized application will be used for smart contracts that are written and placed on the Ethereum marketplace. Dapp will run from a central location or Ethereum nodes.

Blockchain: this is a public ledger that will hold all the transactions done on a cryptocurrency network.

Ether: this is the digital currency used with Ethereum. Ether is referred to as ETH, and one ETH will be worth sixty-five cents in US dollars.

Starting your contract

It is not required, but when you are making your contract, you will be making your own Ethereum node, and it is recommended that you do so, even if you do not use it. As you use a node, you will have the power to connect to the entire Ethereum network. This will include the Ethereum tools of Java, C++, Haskell, and Python.

As of right now, the solidity tool is a programming language which will be the primary programming tool. It will be the ethereum version of JavaScript while using extensions of .se or .sol. A compiler is also going to be required. Ensure that you have the C++ library so that you can have all of the tools that you need to write your contract. Downloading C++ will make it to where you do not have to install solc. There is an alternative that you can use, and that will be a web version that you can find at etherchain.org, or you can use Cosmo. The last thing that you will need is web3.ja; this is an API application that will create dapp. When you have a solidity contract compiled it will be sent to the network. From there, the contract will be recalled with web3.js. So, now, you will have the option of building web applications that will interact with your contracts appropriately.

If you are interested in taking advantage of a framework that already exists, then you will need to use the distributed application framework called Truffle. Truffle will be the recommended choice when it comes to using a basic program that you will understand while allowing for a more significant emphasis to be placed on the individual code. If you do not want to use your own node, then you will need to use blockapps.net. This is another API that will give you the effect of working with a node for testing purposes without having to work with your own node.

There will be specifics in contracts that will vary greatly which means that there will be variables that must be inside of each contract in one form or another. Take for example that if an event occurs then the result will be kept in a log that will create an agreement, but it is not going to affect how the deal acts. At the same time, there is a function that will alter the yes and no state of the contract through the modification of the values that were put into place before the contract was activated. It is because of this function that the move will be moved from one account to another when conditions are met.

The contracts address will be will determine the location of your wallet and if the contract can access your wallet through its unique address which will keep the creator address separate from the wallet address. The next variable will be the size of the contract; the smaller the contract, the better it will run. On top of that, a smart contract will be able to pull data from the oracle by using a public variable that will end up determining if the data needs to be consulted or not and where that information will be coming from.

Other things you will want to consider

As you are writing out contracts, you will need to look at the information that you are working with so that you can make sure that you are storing all the data in the contract so that the terms can be met. You will have a list of items that will end up determining the overall structure of the file that you are modifying. This data is usually going to be stored in a 2-xn mapping sequence. N will be the number of transaction that will be finished along with the specific details that are tied to each transaction.

As you keep the outcome in mind, you will need to include the definition for two different struts. The first one will include information for the person who started the transaction, and that means that the amount of the transaction will be in this strut. The second strut will hold the data for the storage and any information that is needed for mapping the contract as it should be. It is with this strut that you will outline the database that you are working with so that it will automatically label the contracts based on the template that you have created.

Now that you have gotten the template creation out of the way, you will define the functions that will be executed on a regular basis and the prompts that will allow it to carry out the tasks that happen daily. The proper transactions will be sent to the owner of the contract. This idea was first proposed to include the limits that the transaction will add, the contracts fund account, and the stipulations that will apply.

The investor's transactions will be identified by a unique ID that will be stored in the space that has already been set aside for that record, and that has to do with that contract. This area will have been set aside beforehand by the system so that all the results from the contract are kept in the same place. If there is a time limit that is placed on the transaction, then the final deal will be generated. This will then trigger the last action which is often referred to as the suicide action so that the contract cannot be rerun.

It is now that the user will have the opportunity to figure out what they want to do with the funds that they have gotten. These results will then be tested after the contract has been uploaded. A dummy contract will list you in the investor category. You will be required to go through the motions of interacting with the agreement to make sure that it will react in the way that you want it to.

Proper execution

As you saw earlier, Truffle will be the programming process that will make your contract more manageable if you are not comfortable putting all your information together before you test your code first. To help the writing process, Truffle will check the scenarios in the contracts through a Java framework. You should keep in mind that the transaction will take around ten seconds to be verified and that is only if everything in the contract is written correctly. It is vital that you are taking the time into account as you test your code for your contract.

You must have access to the window on the console you are writing your smart contract on, so a new node can be created before the truffle program is opened. When you use the command, it will deploy truffle and cause a spawn to occur in the basics of the init for the smart contract.

The code can be tested through the compiling all the code and checking it for errors and deployment errors.

What happens next

Once the contract's code has been written, it must be formatted so that it can be placed on the blockchain; and you will do this by using an online compiler for solidity found at etherchain.org/solc. At the point in time that your code has been formatted, you will upload the contract by paying the small amount of ether to receive a signature box where you will insert your private key, marking the contract as yours. You are then going to receive the results through an ABI and the blockchain address where the data is now going to be stored for the rest of that contracts life.

After everything has been compiled and updated, the contract needs to be deployed through truffle. You have to open the truffle console and access a new directory by inserting the init command. A new index will be created, and it will place an extension of .sol on the contract. You are now going to enter config/app.son so that you can add the contract to the space that has been provided for it. All you will need to do now is restart your program and run the tesrpc command so that truffle is deployed at the root level. But your contract should be live on the blockchain.

Once the contract has been created, then there has to be a user interface formed that will make it to where you will interact with the arrangement in real time. The dapp will be inside of a database that contains an HTML based front end, and it will be linked directly to the Ethereum platform. If you are using truffle, then the dapp will tie it into a complete network with CDN access. Dapp's UI will be created in a way that is similar to how websites will be designed.

There will be different frameworks that will help in the creation of your dapp so that it is easier for you to deal with. As you saw above, the truffle will be a tool that you will have the option of using, but it is not going to only take that you can use.

You are also going to be able to use Embark; but, the truffle will be the most straightforward development tool that you can use in the creation of your dapp.

When you use truffle, you will be creating a smart contract. But, it is vital to know that there are other options that you can use so that you are making an informed decision as to which application is best for you.

Truffle will be an application that will do most of the work for you when you are working with dapp and smart contracts.

The Embark application will aid in deploying your contracts so that they are available in the JS code of your choice. Embark is also going to make it to where changes will be tracking all of the updates to your contract. Should this occur, embark will redeploy your contract automatically and dapp as needed.

Dapp creation

When dapp is created, you will want to use truffle since it will compile your UI automatically once it has been established. The truffle director has to be labeled app so that the next time that it is run, it will compile the contact information. It will also collect any new changes into the build folder where it can be called upon in case of emergencies with the truffle application.

To get started the directory has to be labeled as an app so that it can find the background images and the JavaScript code that will be tied to the stylesheets and indexes. Depending on what you are needed, you will have the option of adding the code directly into the file that already exists so that you can obtain the front-end UI option which will cause your contract to be up and running in no time. When you open the app.js file, there will be a section that will provide you with a greeting from truffle in the developer console. When you open this console, there will be a list of active commands.

While thinking about commands, you will need to create a function that will have the ability to be accessed every time that a page loads. To do this, you will need a window added by using the code window.onload in the app.js file. If you do this correctly, there will be an assortment of account details that will be shown in the console browser. Finally, you will use a test. Conference.js function so that you can make sure that the output is working as you want it to. Your output should be the balance amount, and this balance will increase after the situation is deferred.

After the app.js has been created along with the index.html so that it will meet your needs, you will test the results with your node or a sample node that will give you results in real time. Note: the results are not going to be prepared relatively quickly. You will want to use the following code so that you can ensure that it will work correctly.

geth --rpc --rpcaddr="0.0.0.0" --rpccorsdomain="*" --mine --unlock='0 1'verbosity=5 --maxpeers=0 --minerthreads='4' --networkid '12345' --genesis testgenesis.json

There will be two new accounts that will be labeled zero and one. You have to understand that you will need both accounts so that a password can be created for each report separately which will end up generating a json test-genesis file that will be founded under alloc on the account where your ether costs are kept. Lately, you will add the results to your truffle application so that you can recompile the contract and deploy the results once more.

There is an option that you will be able to use to generate a UI to use with any dapp that will be created by silent cicero. You can find this application at dapp-builder.metor.com. It is this tool that will be used in creating HTML code you can later modify for your contracts written out on solidity, web3.js or jQuery. It is not going to run as smoothly as you will want it to, but if you are not comfortable with your skill level when it comes to going through the process on their own. The UI that follows will use the same steps like the ones that you have already seen above. If it does not, then there is a secondary version that is usually going to be easier for you to use in finding a solution to your problem.

It is at this point in time that the contract will be written, but you are not going to be done just yet. You have to analyze the agreement. When you examine it, you will ensure that the deal is written correctly and does not need to be tweaked.

Looking at the variables that are located at the top of your contract, they will look like this:

Address public organizer;

Mapping (address => uint) public registrants Paid;

Uint public nonresistant;

Uint public quota;

Address: since this is the first variable in your contract, it will be your wallet address. The address will be set when the constructor which will be called conference (). But, in most cases, the contract will name it the owner.

Uint: you will see that this will be the unsigned integer. There has to be space on the blockchain; this is why you will need to try and keep everything as small as possible.

Public: there will be a variable called from outside of the contract that you write. When working with a private modifier, it will be called on by the agreement. But, if you try and call upon a variable from web3.js then you have to ensure that the variable is set to public.

Mapping and Arrays: there will be varying levels of support for the arrays and mapping like (address => uint) that will be used by solidity. It is also going to write an address out as registrants paid []. These mappings will hold a smaller footprint. Therefore, the mapping will be used in storing the registrant that paid for them so that their funds are available later.

Extra about addresses: this is a client node that will hold information about your account. Whenever you begin your test, then there will be an array of ten addresses that are available.

Your first account will be labeled zero, and it will be the default for any transaction when the state has not already been specified.

Organizer address vs. Contract address: your contract will have its own contract address once it is deployed and this address will be different from the organizer's address. This address will be able to be accessed through your solidity contract. It will be used in your refund ticket function which will be in the contract as address = this;

Suicide, a good thing in Solidity: should any funds be sent to your contract; they will be held by the agreement itself. With the destroy function, the resources will be released to the owner of the deal. Should this not be put into place, then the funds would end up being tied up, and no one would ever be able to access them. So, it is vital that you include a suicide method in your contract so that if your contract dies, you will be able to collect the funds.

However, if you simulate another party of the contract, then you will have the option of using another address that will be different from the accounts array.

Therefore, to buy a ticket, you will have to buy it through this function.

Conference. buy Ticket ({

From: accounts [1], value: some_ticket_price_integer});

Some Function calls can be transactions: functions will be able to change the state of your contract, and these deals will have a specific sender as well as a value that will be placed inside of the curly braces. The funds are then going to be transferred to the wallet's address. So, with solidity, you will have the option of retrieving values through the msg. sender and Ms. Value where the functions for the solidity are stored.

Function buy Ticket () public {

...

registrants Paid [msg. sender] = msg. value;

...

}

Events: events will be optional when you are going through the process to write your contract. Deposits will be set to be sent inside of the agreement to be logged by the virtual machine. But, they are not going to do anything; it is just going to be good practice for you so that you can keep track of all your transactions that have happened already.

Chapter 20: Risks And Money Management

While the benefits and potential of Ethereum is impressive, there are also many disadvantages and risks associated with it. In this chapter, we'll cover some dangers and risks of Ethereum.

Ether is not designed for real world transactions

Ether, the cryptocurrency used for payments on the Ethereum network, is used to pay for computing power to run dApps and smart contracts. Ether is not designed to be used for payments at shops, online, or as an alternative to other real-world payment methods.

Bitcoin and many other cryptocurrencies are more practical forms of payment with many accepted at shops and websites around the world. Using Ether will depend on the popularity of the Ethereum platform and the number of people running dApps and smart contracts.

Ether May Not Increase in Value

Even if the Ethereum platform gains in popularity, there is no guarantee the price of Ether will increase. Many people buying Ether are purchasing it for speculative profit and do not intend to use it for running dApps or smart contracts.

The current price of Ether may be overvalued by traders and speculators. If they sell, this may dramatically decrease the price of Ether.

Even if developers and companies use the Ethereum platform, the price of Ether may not increase if there is a greater supply of Ether created or demand from traders decreases.

New and Unproven Technology

Ethereum is a new technology. While it has a lot of potential, there are still many unknown risks involved with any new technology.

While many companies have joined an alliance to develop the use of Ethereum within their organizations, this is still at the research and feasibility stage. There have been few companies implementing

Ethereum as a replacement for existing systems. There have also been a few mainstream dApps that have become popular.

Many dangers and risks involved with using Ethereum may still be unknown and may not become apparent until Ethereum is being used on a larger scale.

Issues with Smart Contracts
Smart contracts have a lot of benefits; however, there are also significant dangers with using smart contracts.

The biggest case that highlights the flaws in smart contracts is the DAO hack, as mentioned earlier. A badly written smart contract allowed a group of people to exploit the contract and steal more than 50 million dollars.

Thousands of people looked at the code in the DAO smart contract, and most saw no issues. The problems with the smart contract were only realized after they had been exploited.

With a standard legal contract, if there is poorly written wording that allows someone to exploit the contract, this can be taken to court. The disagreement over the wording and intention of a contract can be settled through legal proceedings.

This is not the case with smart contracts. Once a smart contract executes, it can't be argued with or reversed. There are no lawyers or courts to challenge a smart contract.

Dangers of a Giant Worldwide Supercomputer

If you've seen the Terminator movies, you know that Skynet is a worldwide computer network. When Skynet was activated, it detected humanity as a threat and waged war against humans.

The Ethereum network and smart contracts have been compared to Skynet. Ethereum is a worldwide network of computers that are connected, running applications and code that can't be argued with.

In the DAO hack, people exploited a vulnerability in the code of a smart contract to steal more than 50 million dollars. While there are financial risks with smart contracts, there are also much greater security risks.

Smart contracts are being designed for almost everything, including connecting to household appliances, cars, phones, and other electronic systems.

Computer systems are used by almost every modern army today, with some of the most advanced weapons relying heavily on computer technology. Governments are now looking to implement blockchain based technology to replace their existing database systems.

If there was an error in code or the code was badly written, it opens the possibility that the smart contract doesn't run as intended. The contract could then run without being stopped or altered, leading to potentially disastrous consequences, especially if connected to government or military systems.

Hype around dApps
dApps offer many benefits over existing applications. However, throughout history, there have been many technologies with

benefits over existing options but failed to gain mainstream adoption.

Applications, such as Instagram and Facebook, are free; however, they use your personal information to allow advertisers to sell products and services to you. They are free for you to use, but your personal data is the product these companies are selling to other people.

dApps are controlled by the users, not by a centralized company. The users control their privacy and data. In exchange for this control, using the dApp will generally cost money. Uploading a photo, liking a photo, and other actions on a photo sharing dApp may cost a certain amount of money for each action.

Convincing people to pay for a new application that is similar to a more popular free application they use may be difficult. Privacy and the benefits of decentralization may not be compelling enough reasons for people to use dApps over existing applications, especially if it will cost them money to use them.

Ethereum has been evolving. And as it grows, the system improves. Hence, we can expect that there are still some downsides and flaws while it is still in the development process. These problems may stop the users from controlling the platform freely. But understanding the issues will help you big time. It will help you avoid the issues, thus allowing you to continue to use the system smoothly.

Here are some of the problems that cause an issue in the scalability of Ethereum:

The number of users that are on the network

The usage of specialized and unspecialized hardware. Specialized hardware is more powerful than unspecialized hardware. Apparently, the majority of users use unspecialized equipment.

Another problem that you may face with Ethereum is the time stamping. Basically, a block is created every ten minutes. However, when a block is being built every day, that will mean that the system will go slow. Also, if the blocks are constructed too fast, the platform will be overwhelmed, thus causing issues in the performance of the platform.

If you face any of these issues, you need to report it to the developers. These geeks will try their best to fix the problem so that it will not affect other users. Developers are human too. They make mistakes. But, rest assured, just as how we try to improve our ways, we can expect them doing the same. Hence, if you are having problems with ethereum, no matter how small or big the issue is, you must coordinate with the developers.

There will be issues that will come up from time to time. You can help improve the platform of Ethereum if you inform the developers about it. You can also seek help from other users. They may have faced the problem you are having, and they may have solved that.

Chapter 21: Investing In Ethereum

If you've been reading this book as an attempt to understand the ins and outs of Ethereum without necessarily desiring to start using Ethereum, then you may find this chapter particularly useful. This chapter is going to discuss how you can invest in Ethereum as a company, rather than exclusively participate in buying ether and creating Smart Contracts. We will also discuss the growth of Ethereum as a stock, as well as discuss the risk involved when thinking about making an Ethereum investment.

The Fluctuation in Ethereum's Stock Price
When Ethereum was first brought to the stock market, it was valued at $8 a share. Since January of 2017, this $8 figure has risen to an impressive $350 per share at one point. It's important to note that Bitcoin's stock market price was also valued at $2,300 per share in May of 2017. These numbers suggest that investors have a salivating curiosity for the potential of blockchain platforms. While

these prices are certainly impressive, it has not been exclusively an uphill climb for Ethereum's stock. For example, in June 2017 the stock price fell from $319 to just 10 cents in mere seconds, and has also recently been in steady decline since July. Its current market price is set at $199.92.

One of the most critical reasons why blockchain-dominated applications such as Ethereum have seen such a large swing in their stock market prices is because it is still difficult to anticipate whether or not blockchain technology is going to become the way of the future. The people who are investing in Ethereum are doing so rather preemptively, with the hope that blockchain technology will soon dominate the way in which we process contracts and formal documents. For this reason, it cannot be stated outright whether or not you should invest in Ethereum. You will have to make this decision based on your own individual interpretation of the technology and where you think it's ultimately headed.

Reasons Why Investors are Choosing Ethereum

Of course, just because you ultimately have to make this decision on your own, you should still be as informed as possible prior to making this decision. Below are some important reasons why investors are choosing to invest in a technology that still has quite a long way to go in terms of revolutionary impact on the world:

Investors are Choosing to Invest in Ethereum as a Way to Diversify

A good investor should always be looking to diversify his or her portfolio. The cryptocurrency market in general is one that is quite different from any other ones that currently exist. This can be perceived as either a positive or a negative aspect, depending on your outlook. If your glass is half full, you may see the potential for Ethereum to be able to completely change the way in which people interact with certain businesses. The high market share price may also be causing you to think that there is potential for large growth within this sector. On the other hand, what if this high price is only for the short term? What if someone comes along and figures out how to completely break the mathematical algorithms that help blockchain systems to operate? Additionally, it's important to keep in mind that cryptocurrencies are completely unregulated (at least for the time being). This means that the rules are vague, and legality is just as vague. Either way, investing in cryptocurrency will certainly diversify an investor's portfolio.

Cryptocurrencies are Becoming Less Volatile

While the risks that were just presented in the previous section are certainly valid, it's important to understand that cryptocurrencies have been being traded on the stock market for quite some time.

As you can see from the graph above, cryptocurrency stock has also seen a significant rise in price within the last year. The relatively recent shift in stock price within the last year can largely be attributed to investors being unsure about where the cryptocurrency market as a whole is headed. As new problems arise that need to be hashed out, there are plenty of tweaks that still need to be made within the industry as a whole; however, compared to the past, this volatility is arguably less than when Bitcoin and Ethereum first began being traded on the stock market.

Additionally, the volatility that was largely seen in Ethereum's share price in May 2016 has been seen as both a positive and negative occurrence within the investment industry. To some, the volatility is seen as an opportunity to investors who can capitalize on frequent price changes. To others, this fluctuation appears to be a sign that the cryptocurrency market is experiencing a bubble that is sure to pop once demand becomes too high and the new blockchain applications like Bitcoin and Ethereum cannot keep up. A major reason why the optimistic investors believe that Ethereum's share price will level out is because of their desire to switch to the Proof of Stake function that we discussed earlier. With more security and less money being spent on having to secure the network as a whole, it's been surmised by some that Ethereum's share price will level out and become more consistent as well.

This is an Entirely New Investment Opportunity

Lastly, many investors recognize that investors of the past have never had an opportunity such as the one that cryptocurrency and blockchain technology is presenting right now. Unlike other types of products, people seem to be interested in the future of blockchain technology even though the technology itself has yet to be fully developed. This makes Ethereum and other types of blockchain applications a unique type of investment. Who knows, maybe one day investors will look back and scoff at the $190 share price of Ethereum's stock, because it has only grown in price since its tumultuous days.

This chapter should have made you become aware of the fact that blockchain technology has brought a unique situation upon the investment community. If you're someone who is looking to diversify their investment portfolio, then blockchain technology seems to be a perfect way to achieve this goal; however, it's also important to understand the risks that are involved in this relatively new technology type. If blockchain does not take off and become a revolutionary tool of the future, your investment could end up being a total flop. Because the climate surrounding investing in Ethereum often changes, you should be doing constant research and keeping up with the progression of its investment potential as much as you can prior to making the decision to invest your money in this manner. It might be risky now, but one day your decision to invest in Ethereum could end up being your biggest payday to date.

Chapter 22: Should I Invest In Ethereum?

Without a doubt, Ethereum is one of the most promising investments in technology in the last few years. It has proven to be profitable and is growing exponentially. Since Ethereum's birth in 2016, this cryptocurrency has grown above 1100 percent of its initial value. It has literally turned minimum wage earners into millionaires in a short time. Ethereum's growth is also accelerating. If you invested in February, it would have more than tripled by now.

Basing on Ethereum's current rate of growth, it can reach 500 dollars in value in the very near future. Analysts believe that future can be the end of 2017. Long-term projections value Ethereum to thousands of dollars. If you want to invest in this cryptocurrency, you must do it now before it gets even more valuable.

As discussed in a previous chapter, Ethereum, unlike Bitcoin, is more than a digital currency. It's a network that also allows applications to run within which opens up to more possibilities and services that can be offered by retailers.

So what makes Ethereum a worthwhile investment? Here are three of the best reasons:

More Applications Compared to Bitcoin

Overall, digital currencies have a lot of advantages over physical ones. They are secure, effective, and fast. They are also immune to hyperinflation. But one thing that separates digital currencies is the blockchain, its keystone technology.

Bitcoin was the first ever digital currency to use a blockchain network. All the transactions processed in the Bitcoin blockchain are stored in separate blocks. The blocks are then included with others in a chain. The stability of the block is what ensures the security of the whole blockchain and also increases the stability of the currency itself. The blockchain is widely considered to be the perfect ledger. Digitally, that is.

What makes Ethereum more interesting is its use of a far more advanced blockchain than what Bitcoin employs. Ethereum is coded using the Turing-complete language. To be considered Turing-complete, a computer should be able to run any algorithm. Because of this coding language, any script can be executed inside Ethereum. The blockchain used by Ethereum can record transactions faster than Bitcoin can. Bitcoin processes a transaction in 20 minutes. Ethereum can do it in 12 seconds.

A faster network means faster applications and faster processing. The Ethereum blockchain is considered the best network in supporting any program or business. It has the ability to process problems with precision and speed that is rivaled by no other cryptocurrency.

Imagine a typical Honda car with an engine of a supercar. Assuming that there are no technical issues and that the engine runs perfectly on the car. The normal car now runs faster using the same chassis. That is basically what the Turing-complete coding language is doing on the blockchain used by Ethereum.

Because of the speed, accuracy, and flexibility of Ethereum's blockchain, dozens of big companies have invested into it and each one hopes to profit from the network. This brings us to the next top reason why Ethereum is a good investment. It has the backing of large companies.

Fortune 500 Companies Back Ethereum

The Enterprise Ethereum Alliance or the EEA is a major piece of proof that ensures the longevity of the cryptocurrency. It was formed in February of 2017 and consists of multiple Fortune 500 companies. They collaborated on advancing the technology behind the Ethereum network with the ultimate goal of the currency being integrated into their own businesses. The group is joined by business giants such as Microsoft, Intel, BP, J.P. Morgan, and Thomson Reuters.

Companies like these take risks seriously. They are on the top of the game because they make excellent executive decisions and continue to develop business models.

One of the reasons behind Ethereum sharp spike is when the EEA announced the formation of the group to the public. The companies that form the EEA know that there are lots of reasons for Ethereum to be desirable. The cryptocurrency's blockchain is well known for its efficiency and speed, and the Ethereum network has the ability to process smart contracts.

Smart contracts are discussed in detail in another chapter but using an analogy, let's assume that you have this agreement with your cousin: if you hit him, he can hit you back. Now, imagine that this interaction is happening inside the code world. The smart contract determines the transgression (you hitting your cousin) and executes the agreed reaction (your cousin hitting you). Smart contracts are what make things move along efficiently. They are essential to a blockchain network.

Ethereum was developed with the primary goal of making smart contracts work with it efficiently. Before Buterin created

Ethereum, he was involved with blockchain and Bitcoin. The brilliant developer realized that Bitcoin is unable to process smart contracts which is a serious shortcoming.

The Ethereum Virtual Machine or the EVM is hosted on the Ethereum network and it's the one processing the smart contracts. It also makes decisions or charges accordingly. The EVM has the potential to speeding up existing business processes as well increasing business efficiency. No matter what the process is – payment, demand, transaction, etc. The immediate and appropriate response is received for every action taken.

Efficiency translates to money. That is how the EEA trendsetters see it. The CEOs of these companies are known for accurately predicting economic trends. They realize that moving in early into Ethereum helps them position their companies in preparation for the monumental growth of the cryptocurrency. Other investors are taking notice of this. Some of these investors have begun to buy stocks from these early movers in expectation of the benefits to be reaped in long term. Inevitably, Ethereum will be a part of every business.

Financial Institutions Are Incorporating Ethereum

Cryptocurrencies are being adopted by the general public. Ethereum is well positioned and has more potential compared to the others. Ironically, this adoption has started with institutions that are predicted to be destroyed by digital currencies – banks.

Bitcoin is considered to be the banking community's nemesis. It's a known threat to the current monetary system and this is what gives Bitcoin value in the perspective of some individuals.

Ethereum, however, is better suited for banks because it allows them to prosper in the digital world. Bank of America has taken the first step in working with the Ethereum blockchain. It has already launched an application that is based on Ethereum that helps customers secure transactions. This application was developed with Microsoft's help. The goal is to speed up Ethereum technology's adoption into the financial mainstream.

The application translates customer information into packets in the blockchain. Only the appropriate parties involved in the ongoing transaction can access these packets. Information leaks are then prevented because the information can't be emailed out, and this takes out a lot of privacy concerns.

Ethereum is considered to have phenomenal capabilities when it comes to securing financial transactions and this application proves it. Microsoft and Bank of America hope this application will go beyond being a security resource. It has to win the public trust and be the pioneer in this field. Soon, everyday consumers are going to interact with similar applications that are also based in Ethereum. Adoption of the new technology will happen slowly and the consumers won't even know they are already using it.

When the masses buy new releases of MacBooks or iPhones, Apple is guiding them to adopt the new technologies behind the new product. The same thing goes for the Ethereum blockchain. Big companies will slowly incorporate it into their transactions and will be trickled down to customers and clients. Ethereum is considered by these companies to have major advantages over other digital currencies.

Currently, Ethereum alone is being incorporated by large institutions, given that they have had more than a decade to work things out with Bitcoin. This ensures Ethereum's profitability in the long term.

Conclusion

This book is meant to get your toes wet and further your curiosity into the world of Ethereum. It is certainly not all encompassing of everything Ethereum, as you'll also need to learn the specific languages that Ethereum operates on. Make sure you read as much as you can and until you feel absolutely comfortable with CCs, Ethereum, and smart contracts before putting serious amounts of money at risk.

With that, you should have everything you need to enter into the world of cryptocurrencies, transition into Ethereum, start organizing yourself to write contracts and deploying your first smart contract onto the Ethereum platform. You can decide whether generating income by writing contracts that generate goods that you can sell, by buying and selling contracts themselves or fulfilling contracts on your own is how you want to earn your Ether. You may also find that you don't need to withdraw Ether at all, and you end up using it as an alternative to fiat currency. Or, you could simply watch your initial ETH/USD value increase!

www.ingramcontent.com/pod-product-compliance
Lightning Source LLC
LaVergne TN
LVHW011936070526
838202LV00054B/4670